ICE CREAM CONES as Casualties

The History of the Purdue Creamery and Smith Hall

John M. Cleland

Published & distributed by:
John M. Cleland

in association with:
IBJ Book Publishing
41 E. Washington St., Suite 200
Indianapolis, IN 46204
www.ibjbp.com

ISBN 978-1-939550-77-4
First Edition

Library of Congress Control Number: 2017964327

Printed in the United States of America

"John Cleland tells the real story of the people, places, and things that made the Purdue Creamery and its famous ice cream an icon that lives on in the memories of so many Purdue Alumni. This book preserves those cherished memories for future generations!"

Dr. Michael M. Schutz
Assistant Director; Purdue Extension Program Leader, Agriculture and Natural Resources; Professor of Animal Sciences

"Ice Cream Cones as Casualties *is a personal work of art by my friend, Purdue alumni, and longtime Purdue Agriculture supporter, Mr. John Cleland. John's work captures a detailed history of the Purdue University Creamery and its impact on Indiana Agriculture, the Indiana Dairy Industry, and countless Purdue Alumni. Having personally witnessed John's passion and motivation in authoring this book, it is an educational and entertaining read for those who had the good fortune of enjoying a Purdue Creamery ice cream cone and offers those of us who did not, a glimpse into history for a deeper appreciation of this nostalgic, influential Purdue institution."*

Kyle Bymaster
Director of Advancement
Purdue College of Agriculture

"Ice Cream Cones as Casualties *is much more than a detailed, historical account of the Purdue Creamery. It is a case study of the impact land grant universities had and continue to have on United States industry and citizens. John Cleland carefully lays out the evolution of creamery science and technology as seen through the Purdue Creamery and driven by the research, teaching and extension performed at Purdue, and other land grant universities.* Ice Cream Cones as Casualties *is indeed an engaging historical telling of one of Purdue's hallmarks, but it also offers a concise example of the infinitely broader impact that our land grant universities, established by the Morrill Act of 1862, have had on our nation and around the world."*

Dr. Brian Farkas
Department Head and Professor of Food Science
Purdue University

Contents

Acknowledgments

While writing this book, I met many wonderful people who were anxious to tell me their stories and remembrances of the Purdue Creamery and Smith Hall. The encouragement they gave and the assistance they provided made this project both fun and exciting. New friendships were formed while old ones were strengthened. Most everyone said "it was about time" for such a writing to be done concerning such a once-popular location on the Purdue University campus.

In many ways, researching this material was like having the job of a detective. Endeavoring to locate new information within old, dusty files, magazines, attics, and basements kept me on the edge of my chair. Being able to put two clues together and come to a conclusion was always exciting. I want to acknowledge the entire staff of The Virginia Kelly Karnes Archives and Special Collections Research Center for the help, ideas, and resources they provided me. At the top of this list are David Hovde and Neil Harmeyer. Both gentlemen gave me new ideas about where to look and pointed me in the direction of the available resources.

When I said I wanted to write a book about the Purdue Creamery and Smith Hall, Fred Whitford was the very first person I was told to talk to. Fred was always willing to provide ideas, information, and encouragement. Allen Goecker provided most of the information on the Winter Short courses and many of its graduates, for which I am indebted. To my mind, Allen is one of the biggest supporters of Purdue Agriculture. I had known Allen previously, but our friendship was made stronger by this experience. Angela Slocum and Bruce Dexter provided me with information and building construction plans. I appreciate their time and efforts.

David Rodgers, a longtime friend and fellow teacher, reviewed the writing and offered constructive comments. MaryGrace Erickson and I developed a new friendship. She was suggested to me as the person to do the keyboarding, particularly the footnotes. Without MaryGrace, I would still be hunting and pecking on the keyboard.

Special thanks go to my son Brice and my wife, Judy. Both made suggestions and corrections and were extremely patient with me. Judy helped me so much with the computer and much of the keyboarding.

Last, I want to thank Lorraine Albright for her encouragement, friendship, and inspiration in the writing of *Ice Cream Cones as Casualties*.

Introduction

I oftentimes think, young people want to make history, old people just want to write about history. As the years pass and I witness more, I continue to see the importance of preserving historical records with the hope that they will be appreciated by others. So when did I decide I wanted to preserve some of that history myself and write about the Purdue Creamery, Smith Hall, and the people who helped make it a noted location on the Purdue University campus? What initiated this book and what inspired it? Was it inspired by me getting older or by longtime friendships?

In February 2014, Jack Albright called me from his hospital room and asked me to drive to St. Vincent Hospital in Indianapolis to see him as soon as I could get there. Jack was in the last few days of his life and wanted me to finish a handful of tasks that would have been left unfinished if not for the phone call that Saturday morning. He felt these endeavors were important enough to be finished before he passed away. After my arrival, we talked about those tasks, what his wishes were, and how I should go about finishing them for him.

I had known Jack, or at least known who he was, since I was a young 4-H boy entering a show ring with a dairy heifer on the other end of my show halter. My high school FFA advisor, Mr. Simon Deeb, always predicted I would go to Purdue to study agriculture.

As I progressed through high school, Mr. Deeb also suggested that I try out for the Purdue Dairy Judging Team coached by Dr. Albright. During my college career, I achieved that goal as a member of Jack's 1972 dairy judging team. I also was a student in several of Jack's classes. I helped him with small research projects at the Purdue Dairy Farm and at his office in Smith Hall during my senior year, and we remained friends following my graduation.

Jack and his wonderful wife, Lorraine, moved to Carmel in 2010 to live closer to their daughter. This was now just a local telephone call and a short twenty-five minute drive from my house. We called each other on a regular basis and even went out for coffee, lunch, or some social event occasionally. As time passed, our friendship became much more adult and less student-teacher-oriented. However, I must say Jack was the consummate teacher, and I was always willing to be his student.

As we were finishing the conversation in his hospital room about unfinished tasks, he apologized that he never asked me to co-write a published article with him. I told him that was never important to me and that he certainly didn't need to apologize. As a student, I had helped collect some data and do some small chores, but having my name on an article was never on my radar. I could tell it was important to him, however, so I asked him, "What would we have ever written about?"

Looking back, I think Jack had planned this conversation. Unknowingly, I had just taken the bait. He reeled me in faster than a bluegill caught by a young fisherman at his grandfather's farm pond. Jack quickly blurted out, "The Purdue Creamery."

Just as quickly I said, "I can do that." After Jack's death, I jumped to the task. Following some research, I soon had twelve or thirteen pages written and thought I was done. It then sank in that what I had written was either much too long or much too short to get published. So I shortened it to about a thousand words, a much more palatable length for magazine or newspaper publication.

I sought out Tom Campbell of the Purdue Agricultural Communications Department for advice and comments. After making a few changes, he asked if I wanted the article to be used in *Connections Now*, a Purdue College of Agriculture online publication. I was sure Jack was smiling down from above. Jack's wish was met as the article was published in the April 2014 issue of *Connections Now.*

The article was also published in *The Farmer's Exchange*, a weekly farm publication for northern Indiana and southern Michigan; *The Milk Route*, the monthly magazine for The National Association of Milk Bottle Collectors; and the magazine of The Federation of Historic Bottle Collectors. That article now forms the foundation for this book, "*Ice Cream Cones as Casualties.*"

On another level, maybe this book started back in the spring of 1972 with the help of Blaine Crowl, my college academic advisor, mentor, and friend. I was sitting in Blaine's office in Smith Hall when he asked me what I was doing for the rest of the afternoon. My response, "Not a thing," was quickly followed by Blaine saying, "Follow me." We went out to the Smith Hall parking lot and got into his old Chevy pickup truck. A few minutes later, we were at his house northwest of town, next door to the Mt. Zion Methodist Church. As we walked into the kitchen, he asked his wife, Helen, to pour me a cup of coffee and announced he would return from the basement in just a couple of minutes. Sitting at the kitchen table with the cup of coffee in my hands, I looked at a shelf on the kitchen wall that was filled with milk bottles. I had never seen so many in one spot in my life. Just then Blaine returned from the basement with a milk bottle in each hand. He approached and as he handed them to me, he said, "We are starting you a milk bottle collection."

With these two bottles, I began to realize that my study of on-the-farm dairy production as an Animal Science student covered only a small section of the road that milk and dairy products travel from cow to consumer.

Years later I developed a friendship with Ernie Probst. Ernie was a Purdue Dairy Manufacturing graduate and a longtime milk bottle collector. Ernie taught me to appreciate the history behind the dairies. He convinced me to keep a record of any previously unrecorded Indiana dairy I could find and to write down information about that dairy. This culminated in the publication of *Indiana Dairy Listings Book, 3588 Dairies.*

Jack Albright knew of my interest in Indiana dairy history. Thinking back a few years, I recalled that Jack encouraged me to complete that edition of my listings book and complete another book

I was working on about embossed Indiana milk bottles, *The Indiana Embossed Milk Bottle Collectors Guide.* This publication was very frustrating to do because of the amount of labor and man-hours it took. I told Jack about my problems, and in his best teacher voice, he said, "John, this is important to Indiana dairy history. If you don't do it, it will probably never get done."

Those very same words continued to inspire me every time I hit a roadblock trying to finish this book. "I heard you Jack. Thanks for your help."

1
Fond Memories

Decades have passed since the Purdue University Creamery closed on June 30, 1969. It is now time to educate a couple of generations about the dairy delights they have missed and about the facility that helped manufacture them. It is also appropriate to celebrate the one hundred-plus years since the opening of Smith Hall, as it helped to train dairymen and others who have entered its door.[1]

Few people who ever walked into Smith Hall prior to June 30, 1969, escaped the magnetic attraction of a Purdue Creamery ice cream cone or some other delicious dairy treat from the creamery salesroom. Nearly five decades later, these people still smile broadly as they think back to their "good ole days" at Purdue University. From across the campus, the Tippecanoe County, and the entire state, they came to purchase an ice cream cone of unequaled taste.

As you entered the small foyer through the north doors of Smith Hall you were just a few feet from ice cream paradise. Walking forward, crossing the east-west hallway, the creamery salesroom was immediately on the right. Looking back at you through large windows of the salesroom were three-gallon tubs of ice cream filled with chocolate, vanilla, strawberry, butter pecan, black cherry, and many other flavors. Each of these delicious flavors tempted you to give it a try. The smiling salesladies; dressed in white, pressed uniforms were eager to mound high a double dip just for you.

Richard Belcher is a 1956 graduate in Agriculture Economics from Rochester, Indiana. Richard remembers Smith Hall and the creamery salesroom very well. A large smile spreads across his face, and his eyes light up when he recalls taking a young Suzanne Biddle to get an ice cream cone on their very first date. She had chocolate, and he had vanilla. They were married in 1957, and this marriage produced three wonderful children. Richard and Suzanne shared a remarkable life together until Suzanne's passing in 2014.[2]

C. R. "Connie" Miracle, a 1958 Dairy Husbandry graduate remembered the dips of ice cream being the size of "softballs." As our conversation continued, Connie let it be known unapologetically that he had consumed more than his share.[3]

Few people have seen or experienced more Purdue Agricultural history than Mauri Williamson. Mauri retired in 1990 and was the executive director of the Purdue Ag Alumni Association for thirty-seven years. He entered Purdue as a freshman in September 1946 after being discharged from the United States Navy in July. Following graduation, Mauri went home and farmed for three years before taking a job at Purdue with the Agricultural Alumni Association. When asked about his memories of the Purdue Creamery, Mauri nearly shouted, "chocolate ice cream cones!" The salesroom in Smith Hall was one of his favorite places on campus.[4]

Mr. and Mrs. Albert Hendress of Wolcott, Indiana, started dairy farming with Brown Swiss cows in 1946. They sold their milk to the Purdue Creamery, which was first transported to the creamery in cans and later by a bulk tank truck. This truck was owned and driven by George Byers of West Lafayette. On Sundays in the warm summer months, the Hendress Family would attend a local church in Wolcott with a packed picnic lunch, swimming suits, and towels to be stored in the car during the Sunday church service. Immediately after church, Mr. Hendress delivered his four sons and Mrs. Hendress to Lafayette's Columbia Park and the refreshing waters of the park's pool. Following the picnic lunch, everyone enjoyed an invigorating swim, but these adventures played second fiddle to the real Sunday treat. The day ended with a trip to Smith Hall and the Purdue Creamery salesroom for the delicious and cooling ice cream. Steve Hendress recalls how a couple of times his dad drove to the swimming pool in nearby Oxford for the Sunday outing rather than Lafayette, but his father's ears only heard complaints from the four boys as there would be no Purdue Creamery ice cream at the end of the day. Steve also hinted that Mrs. Hendress wasn't all that happy either.[5]

As previously noted, George Byers was a milk hauler, delivering milk to the creamery. He worked these milk routes from 1966 to June 1969 when the creamery closed. The tank capacity for his truck was about 2,500 gallons of locally produced dairy farmers' milk. Besides the milk and Mr. Byers, the truck almost always carried one of the four Byers children.

David Byers remembers how it was nearly a daily argument as to who would ride with Dad on his milk route. David and his three sisters knew there was always ice cream waiting at the end of the day. They would get their special treat right after the milk was emptied into the waiting storage vat inside the creamery and the truck's milk tank was cleaned and sanitized. They often ran into the creamery and talked with Creamery Manager Dusty Rhodes or Don Hockema, another milk hauler, as they impatiently waited for Dad. Sporting a large smile on his face, David said, "The wait was always worth it."[6]

Creamery employee dumping and weighing milk delivered in milk cans.
(Courtesy of Purdue University Archives and Special Collections, Purdue University Libraries)

A couple of exceptions to the memories of ice cream come from two men: Bob Hessong and Bob Peterson. Their memories of the Purdue University Creamery are slightly different. Bob Hessong of Indianapolis was a Dairy Manufacturing graduate of 1958. He fondly recalls his days as a student at Purdue. His two favorite professors were Dr. Fred Babel and Dr. Carlton Parmelee because of their friendliness and excellent teaching styles. While most people will mention the Purdue ice cream as their favorite creamery product, Hessong takes a different path and talks about cheese. The famous Purdue Swiss Style Cheese was his favorite. Cheese classes were at the top of his list as a student as well. Hessong also recalls that his manufacturing classes were small, with only five to eight students in most of them, and no females ever joined the classes.[7] The Purdue Swiss Style Cheese Hessong mentions is still for sale through the Purdue Ag Alumni Association and the royalties from these cheese sales fund agricultural scholarships through the Purdue Ag Alumni Trust Fund. Today the cheese is obtained from and manufactured by a non-university vendor.

Immediately following graduation, Hessong put his degree to work for Hornaday Dairy on the west side of Indianapolis. Bob did "a little of everything" at Hornaday, as his title was "Assistant to the Production Manager." Bob says this meant "cleaning out a lot of clarifiers and washing a lot of bottles." Bob later worked for Borden's of Indianapolis as well.[8]

A current resident of Rochester, Indiana, Bob Peterson's memories of the creamery also take a path other than for its ice cream. Peterson was an Animal Science student from 1948 to 1952. As a student working part-time at the Purdue hog research farm, Peterson made frequent trips to the creamery to pick up excess skim milk. This skim milk was then fed to the hogs. Bob, however, also had his share of ice cream.[9]

Jack Alkire, writer for *The Lafayette Journal & Courier,* wrote about Smith Hall in a 1956 article about the Purdue School of Agriculture:

> *The best known building on the agricultural campus is Smith Hall, the fame of which has reached into every nook and cranny of the campus. Students from the school of engineering, science and home economics often visit this building where they claim is served "the best ice cream in the world."*

> *The ice cream is made by the dairy school, which specializes in turning out exotic flavors, usually a different one for each day's production. The ice cream is sold in the retail room at Smith Hall along with other dairy products produced by the dairy school.*[10]

While Smith Hall may at one point in time have been the best-known building on the south side of the Purdue campus, the history of the Purdue University Creamery predates Smith Hall by more than twenty years and three buildings. In many ways, the Purdue Dairy Department was in its infancy until the Smith Hall construction. Even though outstanding research and teaching had been conducted in other structures and locations, Smith Hall brought about larger and more profound developments. These developments not only brought status and fame to Purdue but also an improvement in the dairy industry in Indiana.

Multiple Purdue Dairy Department heads and several creamery superintendents or managers oversaw the production of millions of pounds of delicious and nutritious dairy foods for the consuming public. Along with manufacturing these products, the creamery also educated and trained hundreds of men and women in the proper, safe, and effective production of dairy products for the consuming public. Many of these graduates have gone on to help feed not only residents within Indiana but also people across the United States and throughout the entire world.

This writing focuses primarily on the dairy manufacturing side of the different creameries of Purdue through its buildings, people, products, experimentation, and education. Through the pages of this book, you will find not only memories of the chocolate or black cherry ice cream cones but also an appreciation and understanding of just a small fraction of Purdue University's rich and outstanding dairy history.

2
The Early Years, When Cream Was King

The known historical information about the Purdue Creamery from the days between the founding of Purdue University and the 1890s is indeed intensely nebulous. Using modern-day dairy product analogy to describe researching this subject is, at best, like looking through a glass of skim milk to read a newspaper. At worst, it could be compared to reading that same newspaper through a gallon of double fudge chocolate ice cream. Either way, the evidence is imprecise and lacking.

On September 16, 1890, the members of the building committee of the Purdue Board of Trustees gave directions to make several improvements on the Agricultural Campus. These work and construction projects were to be paid for out of the Agricultural Experiment Station's special improvement funds. The directive ordered the construction of "a frame building for dairy purposes, the present brick building used for that purpose to be torn down and removed."[1] This "brick building" is the only reference found in regard to a creamery prior to 1890. The location, the description, and the real purpose of this structure remain mysteries.

Professor C. S. Plumb, head of the Purdue University Agricultural

Experiment Station and professor of Animal Husbandry, authorized that a frame dairy building be constructed in 1890 at an approximate cost of $500, apparently from the allotted $2,500 of the special improvement fund money. This wooden building was on the west side of South Russell Street, where the former Livestock Judging Pavilion was later built.[2] (Following its use by the Animal Science Department, the judging pavilion was used by the University Grounds Department for many years. The old judging pavilion still stands and presently has a "Building Construction Management" sign on the door.) This wooden, two-story creamery building, when finished, had a general workroom, a milk settling room for the separation of cream and skim, and even an ice house. Apparently, the machinery and equipment purchased for this building were very simple and similar to what some progressive farmers at that time possessed for their own home farm use.[3]

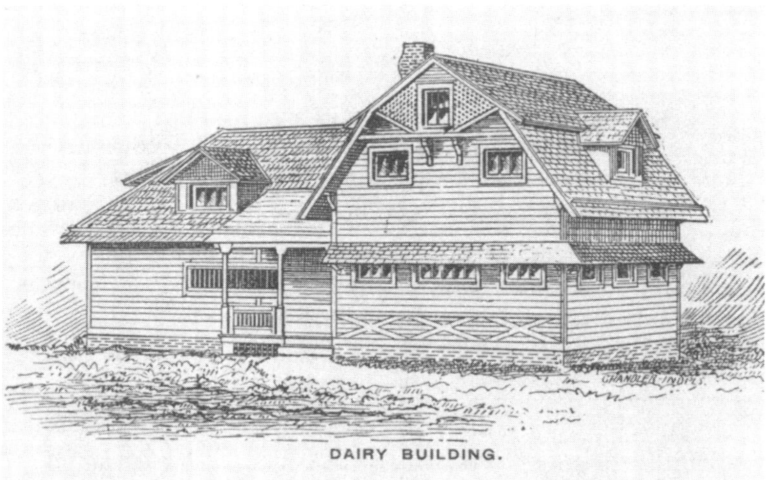

The second Purdue creamery, built on the same site as the first one.
(Courtesy of Purdue University Archives and Special Collections, Purdue University Libraries)

The dairy building contained rooms with several pieces of dairy and creamery equipment. This included two styles of hand-cranked cream separators, a machine-driven Sharpless Standard separator, and a machine-driven Delaval Acme separator. The churn room held two different types of butter churns. The first churn was a Butter Worker, and the other churn was a Power Trunk Churn. Many types of ripeners, aerators, and creamers, along with several miscellaneous items were kept in the settling room. Modern Babcock testing equipment was found in the laboratory of the building. Students also had the opportunity to work in the boiler and engine room, which was needed to power much of the machinery. There were also accommodations for toilets and storage.[4]

William Carroll Latta, Purdue Agriculture Extension pioneer, described the building as being "specially constructed for its purpose" and "a model building of its kind." He also described it as having four rooms: a general worker room, a milk settling room, a cheese room, and an ice house.[5]

It wasn't until 1892 that experimentation by Purdue University staff and students was performed to gain dairy product and creamery production knowledge. Until this time, the only education students received was through demonstrations by the Purdue staff.

The first known manager of the Purdue Creamery was M. L. Fisher, dating to 1895.[6] Fisher was in charge while he was still a student studying agriculture.[7] He worked in the small building where he separated cream and churned butter twice a week. The milk that Fisher churned was from the Purdue Dairy, which at that time consisted of ten Jersey cows. These cows were purchased by Professor Plumb, with the herd located on pasture ground and in a barn where the Lilly Hall of Life Science is today.

M. L. Fisher.
(Courtesy of Purdue University Archives and Special Collections, Purdue University Libraries)

After churning milk into butter, Fisher then had the responsibility of cleaning the equipment, caring for the butter, and delivering the butter to his customers on Saturdays. One of his regular customers was then Purdue University President, James H. Smart. Several other Purdue professors were part of Fisher's Saturday route as well. Fisher made monthly collections of the money owed.[8]

Fisher was apparently a part-time student as well as the caretaker of the creamery. He required several additional years to finish his degree in agriculture because of this work. The 1905 *Debris* (the Purdue University student yearbook) wrote: "Though it required eight or ten years to graduate, he finished."[9]

These eight or ten years were actually in two different stints of time. Fisher was an active student from 1894 to 1896. He apparently then dropped his enrollment and worked at the creamery until 1901 with the opening of Agricultural Hall and a new creamery in its basement. Resuming his education at this time, he finished in 1903 and accepted employment with the university in the teaching of agronomy.[10] Following Fisher's graduation in 1903, he became well known and well written in agronomy, particularly concerning feeds and legumes for dairy cattle. In the summer of 1911, he received his master's degree from the University of Wisconsin.[11] Fisher followed this by getting into the field of agricultural administration as the assistant dean of Purdue Agriculture. He was promoted again in 1925 to replace Dean Stanley Coulter as dean of men for Purdue University.[12]

The next person to become head of the Purdue Creamery was Hubert Everett Van Norman. In 1898, he was employed by the university as superintendent of the university farm and as an instructor in dairying.[13] Van Norman had graduated from the Michigan Agricultural College (today Michigan State University) the previous year in 1897.[14] The Tilsonberg, Ontario, Canada, native came to the United States in 1880 to further his education.[15] He took the job at Purdue for a salary of $600 per year, which was increased to $720 in 1899.[16]

In 1901, Van Norman took over the operations of the creamery from Fisher and began teaching dairy coursework to Winter Course students.[17] He became the head of the entire Purdue Dairy Department in 1902.[18] At this time, the Purdue School of Agriculture began transforming from general agriculture studies into more specialized curricula, though the change would not be completed for a few more years.[19] To illustrate this, in 1900 only one class in dairying was taught. Agricultural students took a broad spectrum approach to general agriculture, with a class or two from each of today's disciplines. The class in dairying was taught in the student's junior year during the eleven-week spring term. At that time, Purdue had a fall term of fourteen weeks, a winter term of twelve weeks, and a spring term of eleven weeks. This dairy class was four credits consisting of four hours of class each week. It included the study of milk, butter, cheese, and methods of handling dairy products, with laboratory instruction. The class was taught by Professor Plumb.[20]

In 1904, the class in dairying was moved to the second semester of the freshman year and was required of all freshmen in agriculture. Seven hours of class and laboratory were required each week. The class was described simply as practice or laboratory work and lectures in farm dairying. It was taught by Professor Van Norman.

A special elective class offered for upperclassmen was also taught by Van Norman. It covered several branches of dairying such as pasteurization, city milk supply, and city and government inspection, with special emphasis on positions in commercial, educational, and governmental work. There was also plenty of practice in scoring and judging of butter.[21]

As department head, Van Norman was responsible for the transition from the wood-structured creamery on the west side of Russell Street to the basement creamery of the newly constructed Agricultural Hall. With this move in location, increased space, and more modern equipment came an increase in coursework offered to potential dairymen. It is interesting to note that, until the fall of 1905, the range of class offerings in the School of Agriculture was quite limited. Every student took every class without a chance for specialization in any area of agriculture. With the start of the school year in September 1905, six areas of intensified studies were offered: horticulture, agronomy, agricultural chemistry, animal husbandry, dairying, and general agricultural.[22]

When Agricultural Hall was constructed on campus in 1901, the creamery was relocated to the east end of its basement. (This building was later named the Entomology Building. Today it is Pfendler Hall.)[23]

> *The Dairy Laboratory [meaning the creamery] occupies half of the basement of Agricultural Hall, and is equipped with factory size cream separators, and combined churns and workers, milk heaters, cream vats, pasteurizers, starter cans, butter printers, cheese making apparatus, and ice cream machinery. The testing room is equipped with hand and steam testers, cream balances,*

> *and apparatus for testing milk, cream, butter, cheese, ice cream, and condensed milk. In the farm dairy room are several forms of churns, butter-workers and prints, sufficient for cream ripening and butter-making on a scale adapted to ordinary farm practice. In the separator room are the leading makes of hand cream separators. Two 7.1 H.P. electric motors generate the power and a 20 H.P. return-flue boiler supplies the laboratories with the necessary steam.*
>
> *The creamery is conducted on a commercial scale throughout the year and is exceptionally valuable for practical instruction and for experimental and research work.*[24]

Professor William C. Latta wrote a letter to the building architect, R. P. Daggett and Company, about naming the rooms of the new Agricultural Hall based on the drawings he had been given. A listing of the creamery and dairy rooms follows:

> *Beginning with the basement there will be under the rear wing a "Farm Machinery room," then at the east end, north east corner, in front "Cheese Room'; next to this toward the center, "Home Dairy" room; next, "Milk Testing" room. At the end of the passage at the east end "Weigh Room," in the south east corner, "Separate Room'; next to this, "Cream Room" then "Churn Room"; the "Cheese curing rooms" are directly under the entrance. The other insulated rooms may be called "Cold or Horticultural storage rooms." The room under the vault "Dairy Store" with refrigerator in the end, fronting on the Hall.*[25]

The section of the preceding sentence where Professor Latta mentions "The room under the vault," referred to the walk-in safe located directly above the creamery on the first floor. This vault can still be seen today after walking up the front stairs of the building and crossing the east-west hallway.

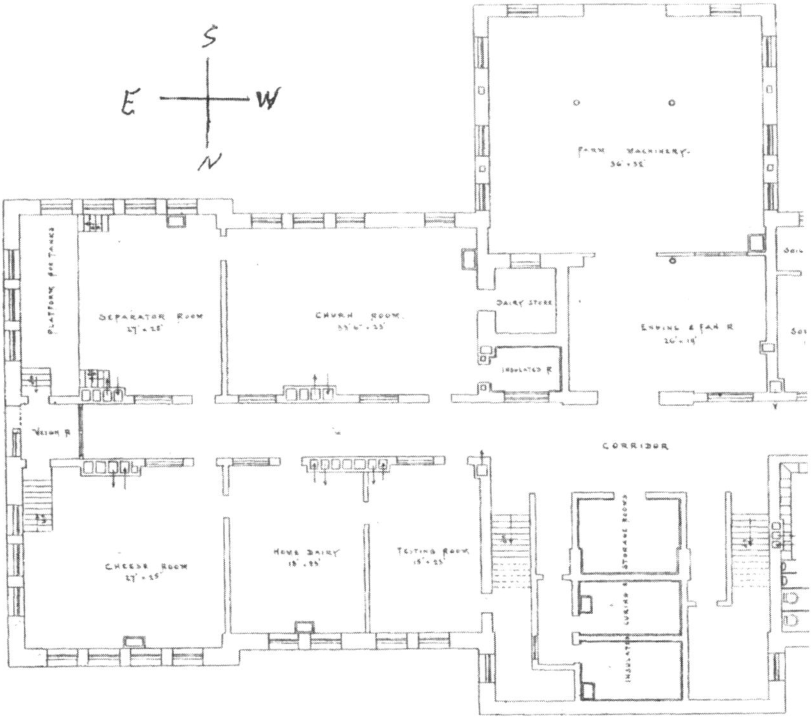

Floor plans for east end of Agricultural Hall basement.
(Courtesy of Purdue University Physical Facilities)

By May 1903, the creamery started processing cream and milk in the Agricultural Hall under the direction of Professor Van Norman. The creamery began to purchase milk and cream from local farmers as it started commercial operations and began to supply dairy products to university students.

Unloading cans of milk at the east end of Agricultural Hall.
(Courtesy of Purdue Animal Sciences Department)

To facilitate increased creamery production, upgraded equipment was purchased. This new machinery included the most up-to-date cream separators, vats, cream ripeners, and combined churns. All of this equipment was needed because Purdue wanted to answer many previously unanswered questions about dairy products.[26] The demand for butter and ice cream was growing, and dairy processors throughout the state were asking questions about increased efficiency and product quality. Additional dairy coursework was offered for students in both two-year and four-year agricultural areas of study.[27]

Van Norman even went to the extent of hiring the additional expertise of H. N. Slater of Fairmount, Minnesota, as a teacher. Slater was also Van Norman's assistant in creamery work and had a reputation as an outstanding butter maker. Following his graduation from the Minnesota Dairy School, Slater stayed and taught classes there. During the winter of 1903, Mr. Slater was hired by Purdue to teach its Winter Dairy courses for creamery production.[28] More precisely, he taught classes in "Creamery Management," "Boilers and Engines," "Starters and Cream Ripeners," and "Butter Scoring."[28] You can find more on Harry Slater in Chapter 9, "Employees of the Creamery."

Professor Van Norman also was given additional duties while in charge of the creamery. He was instructed to take charge of Indiana's dairy exhibits at the 1904 St. Louis World's Fair. During this time, he was named associate professor of Dairying.[30]

Van Norman left Purdue University with his resignation effective on October 31, 1905.[31] He went to work for what today is Penn State University. In later years, he also taught at the University of California. Dr. H. E. Van Norman died July 27, 1938, in Chicago, Illinois, of a heart ailment at the age of 66. He had been the author of many papers and articles about dairying and also authored the book *First Lesson in Dairying*.[32]

In December 1905, Professor O. F. Hunziker joined the faculty as head of the Dairy Department to replace Van Norman. At this time, the entire department consisted of Hunziker, Harry Slater (the butter maker), and a secretary or stenographer.[31] The importance and value of hiring Otto F. Hunziker cannot be emphasized enough. This hiring laid the foundation for Purdue to become one of the great dairy universities of the country. During his eleven years at Purdue, Professor Hunziker proved to be a true visionary for Purdue University and the American dairy industry. The Swiss-born scientist moved to the United States in 1893 and graduated from Cornell University in 1900. A year later, he obtained his master's degree, also from Cornell. Following his move to West Lafayette and Purdue to head up the Dairy Department, he was one of the founding fathers of what today is the American Dairy Science Association. He and seventeen other dairy scientists met in Illinois in 1906 for the initial meeting. He later became the third president of the organization, serving from 1910 to 1911.[34]

O. F. Hunziker.
(Courtesy of Purdue University Archives and Special Collections, Purdue University Libraries)

Hunziker's work brought about the standardization of calibration and measurements of glassware used for the Babcock test. The Babcock test was a method to determine the amount of cream or butterfat in a milk sample. Hunziker was a pioneer in using science to improve the operations of dairymen and milk plant operations. More specifically, he discovered the cause for the fishy taste and tallowness in some butter along with methods for preventing these problems.[35] Hunziker also wrote textbooks that were used by dairy educators across the country.[36]

In 1930, he was named one of the ten "Masterminds of Dairying." This honor was voted on by fifty-six of the leading dairymen of the era. It is believed that this selection committee consisted of members from the American Dairy Science Association. Others selected for the award were Stephen M. Babcock, William D. Hoard, Clarence H. Eckles, Carl G. P. DeLaval, Elmer V. McCollum, Louis Pasteur, Theophilus L. Haecker, Gail Borden, and W. A. Henry.[37] In 1964, nearly 300 leaders of the American dairy industry voted Otto Hunziker as the third most influential person in American dairy history, even ahead of Gail Borden.[38] Hunziker attained not only national fame but also international fame and recognition in the dairy industry. He represented the United States three times at international dairy congresses. The governments of Italy and Switzerland awarded him special recognition and honors. The United States Congress paid special tribute to his service both to this country and to the world for improving the condition of mankind. In 1932, Purdue University presented Professor Hunziker with an honorary degree of Doctor of Science. Even after leaving the university in 1917, Hunziker continued to support and promote Purdue, and he continued to make public speeches and write articles until his death.[39]

In 1905, five classes were taught in dairying, and Professor Hunziker taught all five. A class entitled "Milk" was a required class for all sophomores. "Creamery Butter Making" was a junior-level elective taught the first semester, and "Milk Testing and Cheese Making" was taught as a junior elective the second semester. "City Milk Supply" was taught the first semester as a senior-level elective, and "Miscellaneous Milk Products" was offered as a senior-level second semester class.[40]

At the time of his hiring, Professor Hunziker realized there was a great opportunity for the Purdue Dairy Department to grow and to become more productive in research, student education, and extension outreach to the Indiana dairy industry, particularly the manufacturing sector. In 1906, he headed a successful effort to put together a dairy bacterial laboratory and a chemical laboratory for new research to support the dairy industry.[41]

Purdue was also emphasizing experimentation in the area of butter manufacturing. Starting in early 1906, Purdue spent considerable time and energy in the experimentation of moisture control of butter. Results of this experimentation were published in late 1912. The experiments focused on which factors were and were not under the control of the butter maker, methods of moisture control, and factors related to the accuracy of measuring moisture in butter. The results brought about many new pieces of information and elicited some changes in the way butter was churned.[42]

Cream testing experiments were also run to determine the accuracy of different types of tests and to find the best method of testing.[43] At the time of these cream separation experiments, milk on the farm was not being sold and purchased by volume, as much as it was by cream or butterfat content and cream quantity. This was the period of time when "Cream was King." Factors being tested during these cream separation trials included the richness of the milk, the temperature of the milk, the speed of the separator, the position of the cream screw, and the rate of flow.[44]

Also in 1906, Hunziker hired John DeMoss Jarvis to head up the creamery and support the dairy interests of Purdue and the state of Indiana. He was titled by the Board of Trustees as "Assistant in Dairying." On October 8, 1906, the Board of Trustees gave official approval to the hiring of Jarvis at an annual salary of $1,000.[45] Professor Jarvis was a graduate of the University of Wisconsin and a veteran of the Spanish-American War. Upon his arrival in West Lafayette, he found the creamery to be operating at a loss, but he quickly turned it into a profitable endeavor.[46] The Board of Trustees promptly rewarded him with a $200 raise on June 12, 1907.[47]

In 1907, sixteen students were studying dairying at Purdue.[48] The basement creamery was crowded and was operated entirely by these students.[49] Of the sixteen students enrolled, six had prior experience with farm dairies, while seven possessed experience in creameries. Following their graduation, the university could often help these students find good employment in creameries, milk plants, and depots or on dairy farms, because demand for trained men with dairy expertise was high.[50]

By June 1908, the Purdue Creamery was a profitable and productive business. Production of butter ranged between 4,000 and 5,000 pounds per week.[51] The cream needed for this production came from as far as sixty miles away with up to 6,000 pounds arriving weekly.[52]

By 1911, butter production continued, and a small amount of Cheddar cheese was produced annually. The creamery also manufactured and taught students lessons in production of ice cream and cheese making. An advertisement in the *Purdue Agriculturist* on October 1911 mentioned between 500 and 1,000 pounds of full cream Cheddar being manufactured on an annual basis, along with 18,000 pints of pasteurized sweet cream. This advertisement also mentioned that the creamery's main goal was to give direction and instruction on butter making.[53]

Making cheese in Agricultural Hall.
(Courtesy of Purdue Animal Sciences Department)

Purdue butter was commanding the highest market value based on its outstanding reputation. This output was shipped not only to adjacent towns and cities but also to Indianapolis, Chicago, and as far away as New York City.[54] The New York shipments were sent only during the cooler months of the year, with butter shipped in sixty pound tubs.[55]

The 6,000 pounds of cream Purdue was purchasing every week needed to be paid for. Purdue bought this supply from selected producing farms according to standardized grades based on the cream's quality. Different grading standards were used in different parts of the country. Purdue paid its cream producers based on what was called the "Elgin Grade."[56] The Elgin Board of Trade was formed in 1872. Elgin, Illinois, was a highly productive and influential milk-producing area for products serving the Chicago market. The Elgin Board of Trade produced stable markets for dairy products and also maintained strict grading standards on which the "Elgin Grade" of cream was based. This Elgin Board of Trade was the center of influence for much of Indiana, particularly northern Indiana.[57] Today this is controlled by the various dairy marketing cooperatives and milk marketing boards.

Purdue paid their suppliers of cream about $35,000 annually at the start of the 1920s.[58] The second grade was two cents lower than the first grade. The actual prices paid for this cream have not been ascertained. Checks were sent out twice each month based on the pounds of cream, the percent of fat, the pounds of fat, and the price.

The day the cream was received, it was pasteurized at 185°F and then cooled to 65°F to ripen. A whole-milk starter culture was added and soured by use of lactic acid bacteria. The following day, the cream was cooled and churned until small granules of butter formed, roughly the size of grains of oats or wheat. After the buttermilk was drained away, the butter was washed with water at 65°F. After the working and salting process, the butter was packed into ninety-pound boxes. After they were cooled overnight in a cooler, one pound squares of butter, fifteen squares at a time, were cut by wires. They were then wrapped in parchment paper and placed into cardboard boxes coated with paraffin.

The big butter churn in Agricultural Hall.
(Courtesy of Purdue Animal Sciences Department)

The winter short courses required additional cream, whole milk, and skim milk to meet the educational demands of the students. Special contracts were written with other creameries to supply 2,500 pounds of cream, with similar amounts of whole milk and skim milk.[59] At that time, the Purdue Creamery produced all of its product with equipment from one company, the Creamery Packing Company of Chicago.[60] This company was highly respected throughout the dairy manufacturing world and was thought to be the finest of its kind.

The Dairy Department offered not only winter courses to students but also specialized coursework for experienced butter makers. The first known course ran from December 7 to December 18, 1908. This eleven-day course was not designed to increase the quality of butter being manufactured, but to improve creamery management and cover practical topics such as acquiring butter markets, economic manufacturing, accounting, and others ways to make a creamery a success.[61]

Mr. H. C. Thompson, writing in the *Purdue Agriculturist,* extolled the virtues of a good butter maker. Thompson even thought that the man was far more important than the building that housed the creamery and the equipment put into use. Thompson wrote:

> *With the buttermaker lies a greater part of the success of the creamery; he must be more than an efficient machinist, book-keeper, or what not; he must be an all-around man of business ability and must thoroughly understand his business from floor to roof. The buttermaker that is capable of handling his work, of handling the patrons, that takes an interest in the success of the creamery, is the one that systematizes his work in such a manner that due time is given all phases of his labor. The man that waits till one thing is done before he commences another will make no advancement in the creamery.*[62]

The success of the Purdue Creamery was based on this philosophy of Thompson. The man, the butter maker, is more important than the facilities. The person who can overcome the difficulties is the person who will bring success to the creamery, but the Purdue Creamery was beginning to have its own difficulties.

As previously stated, the dairy laboratories were at the east end of Agricultural Hall, and workable space eventually became a rarity. New machinery and equipment were so closely packed together that it was very difficult for students, instructors, and workers to find a path in which to walk.[63]

Students in a crowded Creamery Butter Making class, 1911.
(Courtesy of Purdue University Animal Sciences Department)

From the two-story frame creamery building on Russell Street in 1890 to the end of the Smith Hall creamery in 1969, a lack of space for work and teaching was always a limiting feature within the Purdue Dairy Department. In an October 1910 letter from Dean John Harrison Skinner to Purdue President Winthrop Stone, Skinner wrote of the building needs of the School of Agriculture. He asked for a new beef cattle barn, a new horse barn, a new veterinary clinic and hospital, and new greenhouses. But with all these perceived needs, he suggested that a new dairy building was by far the greatest and most urgent need. Skinner estimated that a new dairy building would cost $60,000. In his letter he wrote:

> *The equipment for instruction in dairying, although good as far as it goes, is entirely inadequate to take care of the present number of students in these lines. The laboratories are small and are so arranged that it requires a much larger number of instructors to handle the students than it would if we had a suitable building with laboratories of sufficient size and equipment. Not only this but it [is] impossible to give satisfactory instruction with our present facilities.*
>
> *The large increase in the number of agriculture students, 45% in 1910, has over-crowded [sic] the laboratories in Soils and Crops and by erecting a new dairy building, we can relieve this situation.*[64]

In 1909, a new smokestack and a new sewage disposal system were put into place for the creamery in its basement location.[65] Other additions and repairs to the creamery continued over the years. For instance, in the autumn of 1911, a Victor churn, a Whizard cream ripener, a self-emptying starter can, and a brine manufacturer were put in the basement creamery. All of these newly added items were sealed with white enamel. This enamel made them look more appealing and supposedly were more sanitary.[66] The Whizard ripening tank with revolving coils for even heating of cream was part of the equipment used in the pasteurization experiment that Purdue conducted.[67]

In 1908, Purdue opened its new Agriculture Experiment Station Building just east of Agricultural Hall. Today this building is the Agricultural Administration Building. At the time, it was the shining star on the Agricultural Campus; the administration wanted to use it to publicize the work Purdue was doing to benefit the farmers of Indiana. In July 1909, each department wrote a few pages concerning the agricultural work and experimentation it was conducting, which were included in a *Purdue University Bulletin* that was distributed to farmers around the state. The Dairy Husbandry Department goals were (1) to release the most modern information on "The most economic production of sanitary milk" and (2) "to assist in the manufacture of butter, cheese, ice cream, and condensed milk in successfully turning out high quality product."[68]

More specifically, Purdue sent men to examine possible locations and facilities to establish local creameries and processing centers for milk and cream. Purdue also gave assistance and helpful production information to butter makers about manufacturing high-quality butter. Purdue did this because many creameries around the state were not meeting the criteria of butter as defined by rules and regulations of the Internal Revenue Service. Purdue also offered bimonthly butter-scoring contests so butter makers could perfect their skills in scoring and manufacturing butter. These contests were designed so that the creameries and butter makers bore little to no expense.[69]

A class in scoring or grading butter in Agricultural Hall, 1909.
(Courtesy of Purdue Animal Sciences Department)

During his time on Purdue's campus, Professor Jarvis taught dairy manufacturing work. He taught the Winter Course classes and also classes for regular four-year students. He instructed classes in "Creamery Management," "Ice Cream," and "Creamery Buttermaking."[70] Jarvis apparently put into real life what he taught in the classroom, as attested to by a 1908 State Board of Health report in which the Purdue facilities were found to be "in excellent condition."[71]

In 1912, John Jarvis left Purdue. He was seemingly enticed away by the Delaval Company for their educational work. In this new position, he was involved in teaching farmers and creamery owners about the advantages of Delaval dairy equipment, particularly the cream separators that Delaval wanted to sell.[72] The 1912 *Debris* reported a much smaller Agricultural faculty count than in prior years. The entire faculty was back to thirteen because Roberts, Crane, and Jarvis all left Purdue's employment, leaving Hunziker as the remaining member of the dairy instructional staff.[73]

As the calendar switched from 1912 to 1913, one of the most important years in Purdue Dairy Department history was heralded. First of all, additional dairy staff members were quickly hired. By February 1913, the dairy instructional staff comprised Professors Caldwell, Mills, Roberts, Switzer, and McKellip, along with the department head, O. F. Hunziker.[74] Several of these gentlemen were on the staff of the Agriculture Experiment Station but taught classes as well, which exemplified the perpetual close relationship and cooperativeness between the Experiment Station and the School of Agriculture.

The year 1913 also witnessed the graduation of H. R. Lochry from Purdue and the work he began, which was some of the more interesting and unique work experiences of any Purdue graduate, at least in regard to his eventual work location. Lochry was employed by the U.S. Department of Agriculture (USDA) as a dairy manufacturing specialist covering eleven western states.[75] He also helped establish and put into operation the first cheese factory in the Hawaiian Islands. In March 1926, Mr. Lochry took a leave of absence from his USDA employment to put together the equipment and operations of a cheese factory for a man named Mr. Carter. In 1926, Mr. Carter owned one hundred head of Holstein cattle, along with what was thought to be the largest Hereford beef cattle herd in the world and 600,000 acres of land in the territory of Hawaii.[76]

Additionally, because of the increased demand for its dairy products, the Purdue Creamery soon outgrew its basement home. Two hundred local farmers supplied the creamery with whole milk and cream. The Purdue herd also continued to make a contribution. A much larger, more spacious dairy building was required to process these dairy products and to meet needs of the customers.[77] This demand led to the planning and construction of Smith Hall, the building most remembered today as the hub for dairy education and experimentation at Purdue.

3
Smith Hall and Purdue Dairy Legends

In explaining the decision to construct Smith Hall, the *Purdue Agriculturist* wrote, "The Dairy Department has been seriously handicapped the last few years by a lack of space. The new building will be a boost to the department and will enable Professor Hunziker and his associates to do a larger amount and more efficient work."[1]

Construction of Smith Hall began following a gift of $50,000 to Purdue in 1912 by benefactor, banker and farmer William C. Smith of Williamsport, Indiana, who had died in 1911.[2] The money donated was directed toward constructing a building dedicated to instructing students on the principles of modern dairying. The building was named in Smith's honor.

Smith was born quite poor in Warren County. Because of his own hard work, intelligence, and habit of saving his earnings, he accumulated great wealth. He remained a bachelor his entire life and had great admiration for the work Purdue University was doing to improve the living standards of the people involved with Indiana agriculture.

In addition to the $50,000, Smith also gave Purdue 1,500 acres of forest land located in Minnesota. Because the market value of trees was quite low at that time, he recommended that the university hold onto the land for the time being. In addition, Smith provided funding for ten students to attend Purdue University each year tuition-free. These scholarships were open to students of Warren County attending Purdue regardless of their area of study.[3]

The creamery was located at the rear of the new building, with a driveway for milk delivery on the southeast corner of the property. This driveway led to an elevated loading dock that faced east. In 1950, east and west wings were added, greatly increasing the square footage of the structure.[4,5]

The plans for Smith Hall were prepared by the Indianapolis architectural firm R. P. Daggett Company and were approved unanimously by the Board of Trustees.[6] The Daggett Company was a highly prolific firm, designing buildings not only at Purdue but also throughout Indiana. Three generations of the Daggett family were involved in the architectural design of many buildings at Purdue University. Robert Plant Daggett, company founder, designed University Hall in the 1870s. He was followed by his son Robert Frost Daggett, and by his grandson Robert Frost Daggett, Jr.

Besides Smith Hall and University Hall, fifteen other Daggett buildings existed on Purdue's campus. These included the Agricultural Engineering building (1901); the Physics building (1904), which was demolished in 1989; Grissom Hall, originally Civil Engineering (1905); the Agricultural Experiment Station, now Agricultural Administration (1908); the Farm Mechanics building, now Forest Products (1909); the Livestock Pavilion (1909); the Practical Mechanics building, known as Michael Golden Labs (1909); and Lynn Hall of the Veterinary School (1915).

After 1915, only one Daggett building was constructed on the Agricultural side of the campus, the Horse Barn (1918), now Herrick Labs. However, several others were constructed north of State Street. These were Stanley Coulter Hall (1916), the Armory, (1917), Mathews Hall (1921), the Electrical Engineering building (1930), and the Memorial Gymnasium (1936), now HAAS.[7]

Indiana University, Butler University, and DePauw all have Daggett-designed buildings. Several buildings at the Indiana University School of Medicine, including Riley Hospital and Long Hospital, were also designed by the R. P. Daggett Company.[8]

Other Daggett-designed buildings are located at the Indiana State Fairgrounds, Indiana School for the Deaf, and Indiana School for the Blind.

The *Purdue Annual Report for 1912* stated, "A contract has already been let for the construction of this building [Smith Hall] at a cost of $45,464, but this includes no provision for the utilities, apparatus, and furnishings of the building." The annual report went on to say that many of the necessities for the building's use were not included in the original budget. Lacking were plumbing, heating, engines and motors for powering the creamery, and laboratory equipment, along with special laboratory and creamery equipment. Also missing in the budget was furniture for the classrooms and laboratories. Many thought that all of these items would cost another $32,000, whereas less than $5,000 was left from the Smith bequest.[9]

The general contractor for the building was the A. E. Kemmer Company of Lafayette. Kemmer also constructed other notable buildings on Purdue's campus, including the Union Building, the original section of the Ross-Ade Stadium, the Physics Building, and the Purdue Library constructed in 1913.[10] Kemmer, originally from Rushville, Indiana, was a 1902 graduate in Civil Engineering and was president of the student Civil Engineering Society his senior year.

Early phase of construction on Smith Hall and creamery.
(Courtesy of Purdue University Archives and Special Collections, Purdue University Libraries)

Late phase of construction on Smith Hall and creamery.
(Courtesy of Purdue University Archives and Special Collections, Purdue University Libraries)

In February 1913, the *Purdue Agriculturist* wrote, "Smith Hall is being constructed with remarkable rapidity."[11] One month earlier, they reported that the university was hoping for a continuation of good weather so that construction on Smith Hall could be pushed for completion by May 1.[12] In February, the second floor was constructed up to the roofline, and the rear of the building was ready for the roof to be installed. Contracts for the heating system were sent out to potential bidders, and the concrete floors were ready to be poured as soon as the weather permitted. Purdue officials still hoped for completion by May 1.[13]

Smith Hall and the creamery upon completion.
(Courtesy of Purdue Animal Sciences Department)

The weather promoted the speedy construction of Smith Hall, but some credit also should be given to Purdue's managing engineer, Alpha E. Jamison. Jamison oversaw all construction on the campus from 1900 to 1914.[14] He was an 1895 Mechanical Engineering graduate of Purdue and was also later employed as an instructor in that discipline. Jamison authored several textbooks used in engineering studies. He was also considered one of Purdue's finest athletes during that era, participating in football, baseball, and track. In 1915, he was voted the greatest Purdue athlete of the time. Jamison also coached both football and basketball for the Boilermakers. (In 2013, Jamison was selected for membership in the Purdue Athletic Hall of Fame.)[15] In 1915, Jamison left Purdue and went into private business.

Although much of the credit for the on-time completion of Smith Hall should be given to Jamison, the timetable for finishing the work and the managing of the construction budget was on the mind of President Stone as well. At the June 10, 1913, Board of Trustees meeting, President Winthrop E. Stone reported that Smith Hall was within the allocated budget of $50,000 and was nearing completion. He also reported that the $28,000 the Indiana legislature funded for needed equipment would not be made available until December 1913. Stone wanted to move in and use the building as soon as possible and to order the needed items, with the bills to be paid in January of the following year. The building was made available for student use upon their arrival on campus in September.[16]

The 1913 *Debris* described the brand new dairy building as follows:

> *The building is of dark brick and stone with reinforced concrete floors and a tiled roof. It is finished in oak, lighted with electric lights and steam heated. It covers three-eighths of an acre of ground and one half acre of floor space. The building consists of a main front part of two stories 60 x 120 feet; a rear 72 x 180 feet, and a connecting building.*

The rear portion of the building, consisting of the 72-x-180-ft. area, was the new home for the Purdue University Creamery.[17] The *Purdue News,* a publication of the Purdue School of Agriculture, described the creamery as follows:

The creamery annex houses a modern dairy plant operated on a commercial basis and used as a laboratory for students. The creamery building provides rooms for receiving, buttermaking, bottling pasteurized milk, testing, compressors, printing butter, can and bottle washing, and cold storage.[18]

Winter Course students in a cream separator laboratory, March 1917.
(Courtesy of Purdue Animal Sciences Department)

The student creamery laboratory was furnished at a cost of $3,978. Items needed and priced for this room were vats, churns, starter cans, ripeners, pasteurizers, printers, a paraffiner, belts, and shafting. The second most expensive room was the market milk room. It was furnished with vats, separators, a heater, a pump, a clarifier, a pasteurizer, a holder, a cooler, a bottler, a capper, a sterilizer, a bottle washer, and other assorted supplies. All of this equipment was estimated to cost $3,659. Furnishing the lecture rooms was also included in the plans. It is known that Smith Hall had three classrooms and two larger student laboratories.[19] There were also rooms used for stocking, separating cream from milk, farm butter making, cheese production, cheese curing, and packing. Other rooms included a student bacteriological laboratory and incubator room, a receiving room, a starter room, and a cold storage room. An experimental laboratory, offices, and a private laboratory were also present, as well as a tool room and a machinery room. The largest expenses to equip Smith Hall were a refrigerating machine and ice making equipment, which together cost $5,500. Much of the plumbing was not included in the original planning costs and added an estimated $3,765 to the $28,321 estimate for equipping the new structure.

A heating plant for the new dairy building was also not included in the estimated costs for the original construction plans or in the furnishing plans. Professor A. P. Jamison of the University Building Committee estimated this cost to be an additional $3,500.[20] The A. E. Bradbury Company of Lafayette was the contractor for both plumbing and heating.[21]

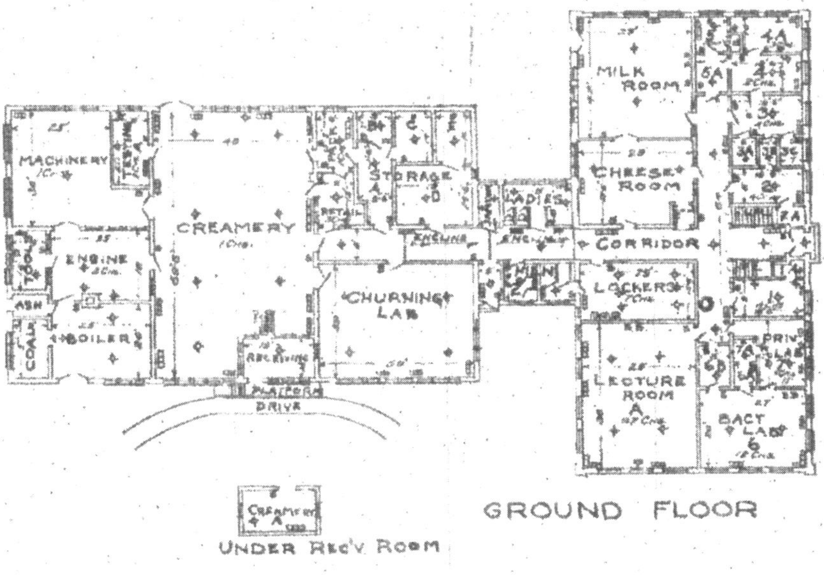

Smith Hall and creamery floor plans, 1922.
(Courtesy of Purdue University Physical Facilities)

As students returned to Purdue for classes in the fall of 1913, Smith Hall was complete and ready for occupancy. Smith Hall was the jewel of not only the Agricultural Campus but also of the entire campus. This new accommodation was the most modern dairy facility in the entire country. Classrooms and laboratories were soon being put to the test by students and faculty alike. The *Purdue Agriculturist* called it "a fine monument to Mr. Smith."

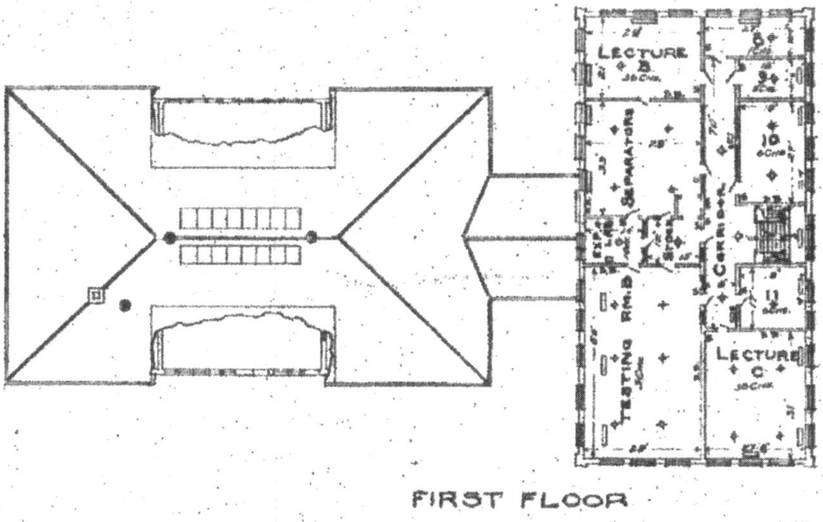

Floor plans for upper level of Smith Hall, 1922.
(Courtesy of Purdue University Physical Facilities)

The land around the building had been worked, and the soil had been seeded for grass, but not everything was complete. The creamery was not yet operational because necessary equipment for its operation still needed to be delivered and installed.[22]

The author of the *Purdue Agriculturist* article went so far as to write, "The new creamery will not be ready for occupancy for some time yet." This statement leaves it open as to when the creamery did actually start production. We do know it was before January 1914 because a photo of Winter Course students working in the new butter room was shown in the April issue of the *Purdue Agriculturist.* The Winter Course started in January and ran for eight weeks.

In an interview article entitled "Quarter Century of Dairy Progress," the *Purdue Agriculturist* (1931), Professor H. W. Gregory reiterated many of the same ideas presented earlier about the planning, construction, and reasons for having a building like Smith Hall. During the interview, he was asked, "Why was the creamery built?" Gregory said the following:

> *The Purdue Creamery was gradually developed to furnish raw materials and facilities for instructional and experimental work. A gift of $50,000 to Purdue by the late William C. Smith of*

Williamsport, Indiana, made possible the erection of Smith Hall in 1913. The Dairy Building now houses practically all of the activities of the Dairy Department; the construction of the Dairy Building was an important step which enabled the expansion program of the Dairy Department to go on. The Creamery now offers materials and facilities for instructional and investigational work in butter, milk, sweet cream, cottage cheese, cultured milk, ice cream, refrigeration, etc. The operation of the creamery is very closely associated with the instructional and investigational work. The creamery is of great importance in laboratory instruction because it is imperative that students receive training in laboratory practice to supplement the theoretical and practical knowledge offered in the class room.

The attributes of Smith Hall placed it at the apex of the dairy education and milk manufacturing industry. The purpose of the building was to train dairy students to return primarily to their small rural Indiana towns and operate the local dairies and creameries. The bacteriological laboratories along with the labs for ice cream and cheese making were on the first floor. The labs for butter making and cream separation were stationed on the second floor. Pasteurization of milk was planned for when the first milk supplies were brought to the creamery.[23]

Winter Course students in Bacteriology lab, March 1917.
(Courtesy of Purdue Animal Sciences Department)

The first manager in Smith Hall was the well-respected Professor Horace C. Mills. He was in charge of the creamery from 1912 to 1917 after having spent the three previous years with the Purdue Agriculture Experiment Station.[4] Mills was directly involved in the output of the creamery and taught classes in creamery management and product manufacturing as well.

In a letter of recommendation for Mills just prior to his 1904 graduation, Professor William C. Latta wrote:

> *I take pleasure in recommending to your consideration Mr. H. C. Mills, a senior who completes his work here the first week in June. He has had the regular instruction in dairy work, and has made a special study of dairy work this past year, and has taken as the subject for his thesis, "Study of the Moisture Content of Butter." In this work he has done a large amount of churning, using the factory size churn. I believe he has the ability to grow into a good man, able to handle considerable responsibility. He is careful—painstaking in his work, quick to "catch on." I have asked him to write you.*

H. C. Mills.
(Courtesy of Purdue University Archives and Special Collections, Purdue University Libraries)

After graduation, Professor Mills went to work as the official tester for the cows at the St. Louis World's Fair. While employed in this prestigious position, he was in charge of testing 120 cows daily for butterfat production.[25] At the conclusion of the World's Fair, he was employed as superintendent of the Dahien Certified Milk Farm of Millbrook, New York. He left New York in 1906 and returned to Indiana to work for Schlosser Bros. of Plymouth. In 1909, he left Schlosser's for employment with the Purdue Agriculture Experiment Station.[26]

Hunziker and Mills, wishing to clarify the purpose of the new creamery and how it tied into the rest of the Dairy Department, wrote:

> *Object of the Purdue Creamery*
>
> *The Purdue Creamery constitutes one of the divisions of the Department of Dairy Husbandry. Its existence and operation are essential to the efficient operation of the Dairy Department; its purpose is to: economically utilize the milk and cream from the University Dairy Herd, consisting of pure-bred, tuberculin tested Jerseys, Holsteins, and Ayrshires, kept in sanitary stalls, fed on properly balanced rations and milked with a milking machine.*
>
> *Make possible the operation of scientific, experimental investigations along all the lines of dairy manufactures, the results of which are distributed to the farmers, dairymen, and creameries of the State.[sic]*
>
> *Another object of the creamery was to allow members of the dairy faculty to stay in continual communication with each other and to discuss and solve the problems of dairy product manufacturing. This would allow the timely and efficient release of down to earth and practical advice to inquiring individuals and businesses.*[27]

Superintendent Mills and Professor Hunziker didn't just sit in the brand new creamery and Smith Hall and admire their new facility. Instead, during the summer of 1914, these two gentlemen traveled around the Midwest visiting creameries and dairy processing plants. During these trips, they tried to assess the processes and procedures that these plants were doing correctly and what needed to be improved. Mills and Hunziker were welcomed warmly wherever they went because they were two of the most respected dairymen in the country.[28]

Mills resigned in 1917 to accept a position with the E.L. Martin Creamery Company in Huntington, Indiana. While on the Purdue staff, he was both superintendent of the creamery and associate professor in Dairy Manufacturing. He published many articles about handling, testing, and pasteurizing cream. Mills also had prominent publications on many aspects of manufacturing butter. In addition, he wrote the yearly reports for the Indiana State Dairy Association from 1912–1915 while serving as the secretary of this organization during those same years. He was a member of the Indiana Manufacturers of Dairy Products, the American Academy of Dairy Science, the American Association for the Advancement of Science, and the American Association of Creamery Butter Manufacturers. Amid the worldwide influenza pandemic, Mills passed away quickly after contracting the flu just prior to Christmas of 1918. The Huntington, Indiana, resident was just thirty-one years of age at the time of his death.[29]

In 1915, a reduced government tariff on foreign butter went into effect. With this lowered tariff, a large quantity of poor-quality butter began to flood the American market. Purdue began a series of experiments to find causes for poor-quality butter. These investigations would later have an impact across the nation and around the world.[30]

In 1915, the creamery produced 204,933 pounds of butter and nearly 23,000 pints of whipping cream, as reported in the annual report of the Agricultural Experimental Station.[31] The creamery manufactured whole milk, cheese, ice cream, whipping cream, butter, and sweet cream. Some of these manufactured products were sold to wholesale buyers who promptly shipped these finished products to Chicago, Philadelphia, and other East Coast cities. Products were also sold locally to restaurants, grocers, and hotels. Skim milk and buttermilk were also made, and most of these products were sold to poultry and hog owners. The raw milk that supplied the creamery came from both local herds of cattle and the Purdue herd. During 1915, 741 different suppliers sold milk and cream to the creamery.[32]

Butter was the main product produced at nearly all creameries in the state and the country, and the Purdue Creamery was no exception. This is evidenced by the butter production in 1915. Because of the lack of refrigeration on farms, the large quantities of raw milk that were transported to the creamery were difficult to handle, keep cool, and prevent from spoiling. In addition, the skim portion of the milk was of little monetary value. Skim milk was often fed to poultry and hogs. The real demand was for cream, which was used in butter making, ice cream manufacturing, and the production of whipping cream.

Because whole raw milk was made of approximately 4% cream or butterfat, the transport of only the cream required very little space and was much easier to move from the farm to the processor. Many farmers owned cream separators that allowed them to separate the cream from the skim themselves. "Ripening," or letting the temperature rise, is desired in butter making. By doing this, a bacterium lowers the pH in the liquid. Butter makers monitor the pH of the cream to know when to begin churning.[33]

Butter production was so important that O. F. Hunziker and H. C. Mills wrote:

> *Good butter is one of the essentials in the diet of young and old, not only for its food value, but also because of its dietetic effect. Bread, "the staff of life," loses much of its value and attractiveness without its faithful companion, good butter.*[34]

Purdue took great pride in its butter and its butter making. The following is taken from the *Purdue Agriculturist* describing the process of manufacturing butter using Purdue's preferred method in 1916.

> *The grading and weighing having been completed, the cream is allowed to flow into storage vats where it remains until the day's supply has been brought in, usually not later than 3:00 P. M. At this time it is allowed to run into two large vat pasteurizers (one for 1st grade, the other for 2nd) where, after the acidity test has been made and the proper amount of neutralizer* $Ca(OH)_2$ *added to reduce the acidity to 3%, it is pasteurized by steam heat to 145° F for twenty minutes. At the end of the twenty-minute period, cold brine is passed continuously through the pasteurizer coils until the temperature is lowered to 45° F. The cream, being in a suitable condition to leave, is not molested until the following morning. As soon as it is possible in the morning, starter is added—the amount usually being about 15% of the volume of the cream. The starter, a pure culture of Lactic acid bacteria, imparts flavor and aroma to the butter.*
>
> *The temperature at churning varies with the different conditions but usually the churn temperature is 50° F and 56° F in winter. The cream is transferred to the churn by means of a steam injector.*
>
> *Before starting the churning, coloring is added at the rate of from 3/4 to 1 1/4 oz. per 100 lbs. of butter fat, depending on the season of the year. The motive power is of course steam as this is the most uniform. The time required to churn depends upon*

many different factors. The fullness of the churn, viscosity and temperature rank among the most important ones. After having churned until the butter grains reach the size of corn grains, the process is considered complete.

The buttermilk is drawn off by means of a suction pump which conveys it into storage vats, where it is held until sold to the patrons and farmers.

The butter is washed twice with water at a temperature of 2° below that of the buttermilk. Six revolutions or more of a churn on high speed constitutes a washing. At the end of the washing process, salt is applied uniformly over the bulk of the butter at the rate of 3/4 to 1 3/4 ounces per pound of butter fat.

Working of the butter depends upon several factors, moisture content, texture and body being the most important. The average number of revolutions on the slow speed usually lies between 16 and 20 revolutions.

The working process having been completed, a moisture test is made. The legal maximum requirement is 16%, the average moisture incorporated being about 15.8% which gives a fairly good overrun.

If the product is to be put on the market in one pound prints, it is put in 90-pound boxes, but if sold in bulk form it is put into 60-pound tubs, which have been paraffined and lined. The total weight of the butter is gotten and from this and the pounds of butter fat churned the overrun is calculated.

The butter is put into cold storage rooms until sold unless it is to be sold in pound prints. The printing requires much time, however. Much labor and time are saved by the use of special printing equipment. Each pound is weighed, wrapped in specially prepared paper and put in one-pound cartons. These cartons are put in 50-pound boxes which are put into cold storage rooms awaiting sale.

The manufactured product is sold to grocers, hotels, restaurants and wholesale buyers on the large markets of Chicago, Philadelphia and New York.[35]

The word "overrun" found in the previous article, when used in regard to manufacturing of butter, is simply the weight of the butter produced as compared to the weight of the cream a butter maker started with. The weight of the butter is always greater than the weight of the cream. Overrun is expressed as a percentage.

The term "butter print," or just "print," also referred to in the previous article is any quantity of butter wrapped in wax or greaseproof paper. Today this usually means a rectangular one-pound block of butter or a four-ounce stick of butter. In very early butter making days, a special cloth might have been used instead of paper. Often these would be collected after use, washed, and reused.

One pound cartons of butter.
(Courtesy of Purdue Animal Sciences Department)

Besides working with high-quality cream, the Purdue butter makers and creamery staff believed that knowing and controlling the temperatures throughout the butter making process were the key essentials in making quality butter. This philosophy was expounded as early as 1908 in an article found in the *Purdue Agriculturist*:

> *A good thermometer should be a part of the equipment of every dairy farm. To the buttermaker it is invaluable, whether he makes his butter in a creamery or in a dairy. The fact that so few buttermakers use the thermometer accounts, in a large measure, for the poor quality of butter produced in the average dairy. The importance of the thermometer is obvious. So much depends on the temperature at which the cream is ripened and churned. This*

brings us to the question, "What is cream ripening?" Cream ripening is the treatment cream receives from the time it is set to ripen until it is put into the churn, including the chemical, biological and physical changes which it undergoes during that time. During the ripening period cream undergoes a process of fermentation caused by certain acid-producing germs. This gives to the butter a desirable flavor and aroma.

The temperature at which these germs develop best to the exclusion of undesirable germs is at from 60 to 70 degrees Fhr. Cream ripened at a higher temperature than the above is liable to result in butter of poor quality and flavor.

The proper temperature for churning is about 50 to 52 degrees in summer and 54 to 56 degrees in winter. If the cream is churned at too low a temperature the process of churning will be long: the cream becomes viscous and adheres to the sides of the churn. Then the granules of butter will be so firm that they will take up the salt with difficulty, and when the butter is being worked a large amount of water is expressed, causing a small overrun.

On the other hand, if the temperature of the cream is too high at the time of churning, the butter will come quickly, but it will be in soft lumps instead of flaky granules, as it should be. This gives the butter a greasy texture and prevents the expulsion of buttermilk. Butter churned in this manner will deteriorate rapidly, owing to the curd and sugar, which are retained. Churning at high temperature also causes excessive loss of water in the buttermilk.

The temperature of the wash water, too, should be watched closely. Too warm wash water makes butter greasy and sticky; too cold wash water has a tendency to make it tallow, crumbly and develop mottle. The temperature of the wash water should not vary more than one or two degrees from the churning temperature.

A dairy thermometer can be had at a small cost and pays for itself in the superior results that are produced through its intelligent use.[36]

A two-line story in the *Purdue Exponent* in late January 1921 again emphasized the importance of butter production by the creamery: "The big churn at the Purdue Creamery was undergoing a general overhauling yesterday. Owing to the fact that it was out of commission yesterday, two runs will be made today."[37] Two runs on one day showed the value and the demand for butter and indicated that production should not be stopped or slowed down.

Within just a couple of years, Hunziker began to realize that the $28,000 allotted for equipment was totally inadequate. Even at the opening of Smith Hall things were tight for the size of the student enrollment in dairying. In a March 20, 1916, letter to Dean Skinner discussing the upcoming annual budget, Hunziker wrote:

> *We were able to get along under the pinched conditions with the limited number of students, but now that the student enrollment in Dairying has doubled in number and additional courses have been added to our work, the hand to mouth type of equipment is beginning to avenge itself and we have got to provide adequately if instruction is to be satisfactory.*[38]

A front page *Purdue Exponent* article dated December 7, 1921, mentions the demand and output of the butter produced by the Purdue Creamery. The demand it mentions, however, is by the local community.

> *As evidence of the high quality and attractiveness of butter made by the Purdue Creamery, the people of Lafayette are now consuming the entire quantity of butter being manufactured by the local plant. About seven hundred pounds are being made daily and sold for local consumption, and the amount now in the storerooms is less than 800 pounds. Formerly half of the butter manufactured was being shipped to other cities.*[39]

Five years later, Purdue once again was selling butter to East Coast markets. In a June 26, 1926, letter to Dean Skinner, Professor Gregory requested permission to visit the butter markets of Boston, New York, and Philadelphia, as well as the Department of Agriculture in Washington, D.C. At this time, Purdue was selling more than 35,000 pounds of butter to Boston and 16,000 pounds to Philadelphia annually. Professor Gregory thought it was in Purdue's best interest to visit these large butter markets each year.

Gregory also wanted to stop in Washington, D.C. to investigate what was new and upcoming in relation to federal government butter inspections, and he wanted to see the work being done by the Bureau of Dairying. The government was in charge of inspecting the Boston, Philadelphia, and New York butter markets. By visiting Washington, D.C., Gregory tried to stay a step ahead of any possible trouble and to maintain sales of Purdue butter to these three large metropolitan markets.[40]

Other products besides butter were also manufactured and sold during this era. Although butter was the main output, butter making by-products and other nonrelated dairy products also made their mark. The December 7 *Exponent* article continues and gives a brief account of those as well.

> *The selling of sweet pasteurized cream is also carried on quite extensively. Sweet milk and cream from the Purdue Dairy herd together with that received from several dairy farmers make up the supply of this product. The whole milk having been weighed separated by means of a steam turbine separator.[sic] A sample is taken and tested for butter fat so as to make standardization possible. The product is then pasteurized in the usual manner, cooled down and standardized so as to contain 36 to 38% butter fat. It is then put up in pint cans and retails at 25 cents per pint. The demand often exceeds the supply of this product.*
>
> *Only a small amount of cheese and ice cream is sold as these products are made by the students in their laboratory work at certain times of the year.*
>
> *Some skimmed milk is sold but the most of it is used, after having been pasteurized at high temperature, as a media to grow pure cultures of Lactic acid organisms which make up the starter.*
>
> *The buttermilk is sold to patrons in the vicinity of the creamery, to the poultry farmers and to livestock farmers. The fat left in this product is only about 10%.*[41]

In 1926, then Department Head H. W. Gregory, and a gentleman named Mr. Clifford B. Breeden entered into an agreement concerning disposal of the buttermilk. Mr. Breeden purchased 150 gallons of buttermilk a day the first week and later increased the amount to 200 gallons per day. The cost was set at two cents per gallon, and Breeden notified the creamery in advance when he was done feeding his livestock with buttermilk.[42] Breeden was the herdsman on the Purdue Swine Farm for thirty-five years and fed this buttermilk to hogs. This was a common practice during that era. Mr. Breeden retired in 1959 after his thirty-five years of service to Purdue.[43]

The power to perform tasks within the creamery was provided by two 20 HP (horsepower) engines, which powered a 10 ton ice machine and the churns. A 55 HP boiler provided the power for the rest of the plant. Two large ice boxes and a brine tank were cooled by a large compressor.[44]

Smith Hall and the creamery were home to one of the early pasteurizing units in Indiana. The milk used to produce butter was heated enough to kill bacteria and disease-causing agents. This was done not only to increase the shelf life of the product but also to ensure the health and well-being of the consumer. Purdue began pasteurizing milk and cream beginning in the 1890s in the wooden dairy building on Russell Street. At that time, Purdue simply used a water heater to pasteurize milk. The hot water was put into a large vat or trough and the milk cans were placed in the water until the milk and cream reached the proper temperature.[45] Techniques and equipment improved dramatically when the basement creamery was constructed and started operating in Agricultural Hall. The effectiveness of pasteurization continued to improve when Smith Hall and the Purdue Creamery opened.

Today we know that pasteurization is vitally important in achieving the goals of longer shelf life and a bacteria-free product, but previous dairymen were unaware of this fact. For many years, some milk producers and medical personnel alike resisted the idea of pasteurization. They feared pasteurization destroyed the health-giving aspects of the milk and interfered with the treatment of their patients.

A rubbing of an embossed milk bottle and a pyroglazed milk bottle touting the benefits of unpasteurized milk.
(Courtesy of John Cleland)

Besides the pasteurizing of the milk and cream, constantly washing equipment and scrubbing and sanitizing exposed surfaces and utensils helped to keep the dairy products clean, disease free, and wholesome. Hunziker and Mills used the phrase, "Cleanliness First." These were the first and last words at the Purdue Creamery. All employees and students wore white clothes, and most wore white hats. The dominant color was also white because it allowed the detection of soil and uncleanliness. This helped improve sanitation, thus ensuring a higher quality product.[46]

It soon became evident that the Purdue Creamery could manufacture products of quality as well as high quantity in the new facility. In 1916, Purdue won the gold medal for the milk and cream exhibit at the prestigious Panama Pacific Exposition. This exposition was a World's Fair of agriculture. Most states within the United States and many other countries around the world sent their finest livestock and farm products to be judged. Indiana's effort at this event was organized by the Purdue School of Agriculture under the direction of the governor and state legislature. Dean Skinner oversaw and headed the effort. With over 2,500 entries, the dairy products contest was the most extensive and largest ever staged at that time. Purdue's score was 95.5, the highest in the College Experiment Station class.[47] An entry from the state of Washington placed second just two points behind Purdue. An entry from California finished third, five points behind Purdue.[48]

Purdue's performance was particularly impressive because the award-winning sample of Purdue cream had to travel 2,500 miles from West Lafayette, which meant that Purdue's entry was transported by train for eight days prior to reaching its destination, San Francisco. Yet Purdue still defeated entries from much closer locations.

Professor H. B. Switzer, who taught the class "Market Milk," was in charge of preparing Purdue's entry to this competition, with the help of Mr. O. H. Anderson. Prior to shipping the cream, milk from individual cows from the Purdue Dairy herd was tested on a regular basis for three weeks. This was done to check these cows for correct milk flavor and low bacteria counts, which allowed selection of the highest quality milk.

All equipment was specially sterilized and protected from the surrounding air, which might have contained dust, dirt, or bacteria. The flanks and udders of the chosen cows were clipped and washed. Each of these selected cows was also brushed prior to milking. The udder was washed with a disinfecting solution to help keep bacteria counts low. The cows were milked by hand and a special milk bucket with a very limited opening at the top was used. The person milking the cow had to direct the stream of milk through this small opening with great accuracy. The cream separator used to remove the cream from the skim was steamed for sixty minutes to sterilize it. The bottles were sterilized prior to filling and then were filled with cream flowing directly from the separator. The bottles were immediately capped and sealed with paraffin and quickly immersed in crushed ice.

A milk can of unique design was used to ship the bottles of cream to the Panama Pacific Exhibition. This can had two layers of metal with cork placed in the middle to act as insulation. The cooled milk bottles were placed in the can and surrounded by ice. The can was then sealed with paraffin and placed in a special cardboard box that was also sealed with paraffin. Upon arrival in San Francisco, ice was still found in the can. This is all the more impressive because it was shipped during the hot weather of summertime.[49]

Purdue also had an entry in the market milk division and used similar techniques for handling and shipping the milk. This entry received a certificate of merit.[50]

None of these outstanding accomplishments happened by accident. Planning and preparation had started in 1912 when the creamery entered and won a similar contest, called the International Dairy Show, in Milwaukee, Wisconsin. Purdue scored 98.3 out of 100 in market milk and 97.9 in market cream. These scores were the top of each category out of ten agricultural colleges that participated, including such notable dairy colleges as Wisconsin, Minnesota, and Ohio State.

Many of the cleanliness techniques and sanitizing procedures used in 1916 had been perfected in 1912. The cow's milk was pretested for bacteria to see which cows would be used. Washing and clipping the cows, sanitizing udders, and sterilizing all utensils and equipment prior to use had all been done before. Even the hands of the men milking the cows were washed just prior to hand milking each cow. The bottles were precooled prior to filling them.

In the 1912 competition, Jersey cows were used because it was thought that they had superior flavored milk over the Holstein and Ayrshire cows, which Purdue also had in its dairy herd. Butterfat had to be at least 4% in the market milk. The Purdue Jersey milk averaged 5% fat.

Four one-quart bottles of milk and four one-pint bottles of cream were entered in this competition. The bottles were specially fitted into a metal holder and placed in a small wooden box filled with ice. This box was placed into a larger wooden box with hay stuffed between the boxes for insulation. The larger box was filled with ice. The cows were milked October 15, the milk was shipped October 16, and judging took place between October 20 and 25. Bacterial counts taken before shipping and at the time of judging showed that very little bacterial reproduction had occurred. This was a very important lesson about keeping dairy products cold to maintain their high quality. This project was carried out by Purdue students and supervised by Professor Hunziker and Mr. H. B. Switzer.[51]

4
Women In The Dairy, Winter and Short Courses

A Course For Women In Dairying

When the United States entered World War I in 1917, labor on farms and in factories across the country became scarce. This was also the case in creameries, cheese plants, ice cream plants, and other dairy product processing centers. Enrollment in universities also took a nose dive. This was true for the four-year programs and Winter Course at Purdue.

During this time, the Purdue Dairy Department used a new and creative method to boost enrollment. The university sought women to enroll in a special one-month-long Dairy Manufacturing course. The class was designed to run from April 8, 1918, to May 4, 1918, so Purdue printed a special advertising flyer with emphasis on women's patriotic spirit. The "Win the War Course," as it was titled, emphasized the scarcity of qualified workers in factories and dairies that processed and manufactured dairy products. This approach to involve women happened twenty-four years and one world war earlier than the famous "Rosie the Riveter" campaign of World War II. Both classroom and laboratory experiences were offered to women. The following is a list of the classes, a short description, and the class time spent each week.

1. *Milk Testing — To train workers to meet the state requirements of Indiana Milk and Cream Tester's Examination. Emphasis was on the Babcock test. 2 lectures and 2 hours laboratory per week.*
2. *Testing Dairy Products — To train in acid test and moisture test for milk, salt test and curd test for butter. Also detection of adulteration of milk and preservatives. 2 lectures and 2 hours laboratory per week.*
3. *Soft Cheese Making — To train in the making of many kinds of soft cheeses from whole milk, skim milk, and buttermilk. 2 lectures and 2 hours laboratory per week.*
4. *Ice Cream Making — To provide practical knowledge and economical methods of ice cream manufacturing. 1 lecture and 1 laboratory per week.*
5. *Majonnier Test — To train in the use of the Majonnier apparatus. Used by the larger plants in manufacturing condensed milk and ice cream. 2 lectures and 2 hours laboratory per week.*
6. *Dairy Bacteriology — To teach sanitation throughout the dairy industry. 2 hours per week.*
7. *General Dairying — To train in farm production. 2 lectures.*
8. *Food Production — To address general problems of food production. 1 hour per week.*
9. *Convocation — A general lecture each Thursday in Fowler Hall. Speakers were men of national importance.*

Enrollment was open to all women with at least an eighth grade education. No admission test was given. Applicants needed a personal reference from a pastor, teacher, or public official attesting to the person's moral character, work ethic, and ability to study and learn. Tuition was free for residents of Indiana, and out-of-state students were charged $10.

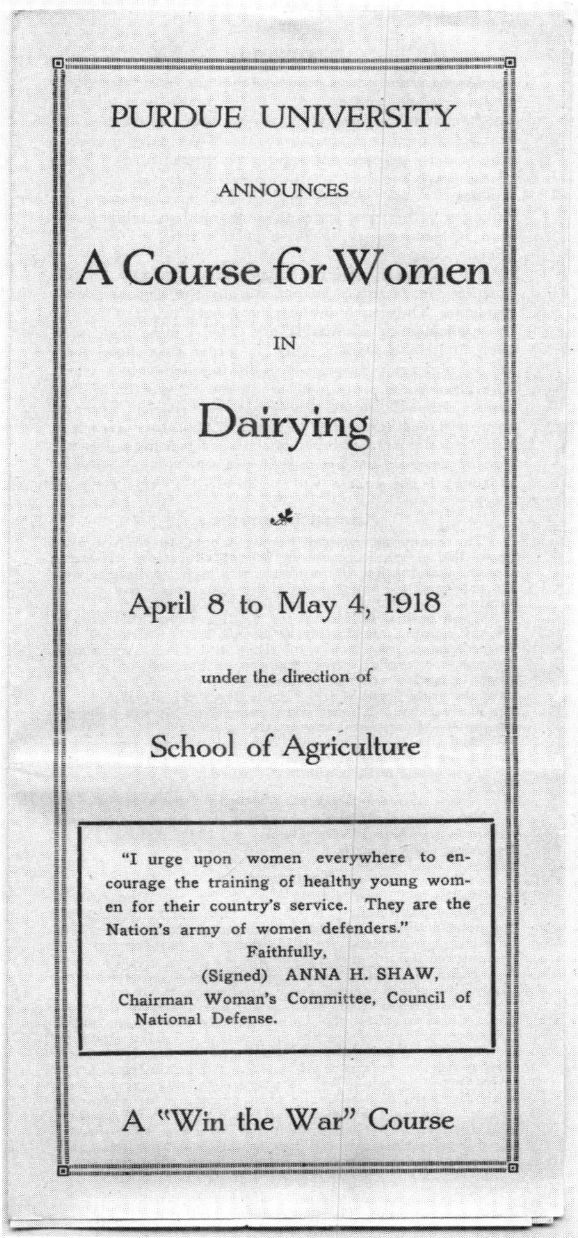

PURDUE UNIVERSITY

ANNOUNCES

A Course for Women

IN

Dairying

April 8 to May 4, 1918

under the direction of

School of Agriculture

"I urge upon women everywhere to encourage the training of healthy young women for their country's service. They are the Nation's army of women defenders."

Faithfully,

(Signed) ANNA H. SHAW,

Chairman Woman's Committee, Council of National Defense.

A "Win the War" Course

Front cover of the "Win the War" course pamphlet.

(Courtesy of Purdue University Archives and Special Collections, Purdue University Libraries)

Purdue estimated expenses for the female students to be about $50 for the four-week course. Room rent was $8 to $10, board $18 to $25, incidentals and laboratory fees $5, books $5, and laundry $5. Proper housing in the immediate neighborhood was arranged through the university.

Employment was not guaranteed by Purdue, but it did promise to notify potential employers of the availability of these trained dairy workers. The instructors were from Purdue's regular faculty, the same gentlemen who taught the classes for the regular four-year students and the Winter Short Course students.[1]

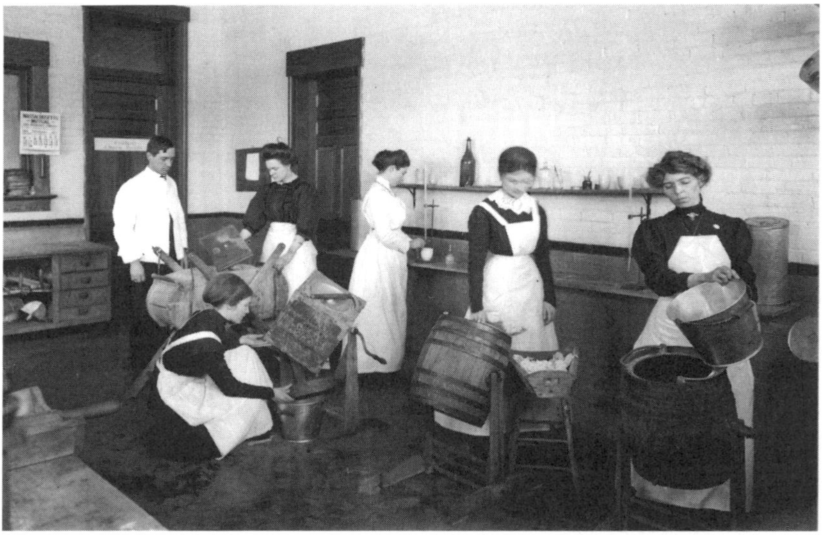

Winter Course women in Household Economics curriculum enrolled in Home Butter Making class.
(Courtesy of Purdue Animal Sciences Department)

This special women's class took place, but no names, enrollment numbers, or number of graduates has been found.[2]

The following was written by Bernard Sobel in April of 1918 for the *Purdue Agriculturist* concerning this special women's course.

THE DAIRY SHORT COURSE FOR WOMEN

By Bernard Sobel

University News Bureau Purdue University

"Where are you going, my pretty maid?

'Sir,' she said,

'Sir,' she said,

'I'm going a-milking, sir,' she said."

She would not say this any longer. Times and customs have changed. The course in Dairying for women now in session has revolutionized matters in the world of pretty dairy maids. Indeed, if you happen to be walking through the campus any bright afternoon these days and should happen to meet any attractive young lady, and you have the courage or temerity to ask where she is going she would answer: "I am going to inspect a large creamery or a cheese plant or a milk condenser."

Now this is how such a strange state of affairs has come about. The present war and its insistent call for food has made it the patriotic duty of every citizen to increase the food production wherever and however possible. The School of Agriculture of Purdue University has felt this responsibility from the first and has directed every energy toward increasing food production. The latest activity has been the creation of a course for women in Dairying, said to be the first of its kind in the history of the United States. The aims of this course have been very simple and very direct. In the wider aspect however, they have great patriotic significance and great commercial possibilities. As a result of the war thousands of men have gone to the warfront. They have left their work behind them. This work in many cases must be completed in order that the army and the allies [are] to be fed. This course intends to prepare women to accomplish this task. In announcing such a plan, the School of Agriculture is, in a sense, defying tradition. In the past, it was thought that women were capable of working at the domestic side of the dairy industry only. This was undoubtedly a misconception, for there are many places that are entirely within the scope of women and certain phases of it which she is particularly well qualified to master. The operation of the Mojonnier tester is a delicate and precise one, but women are, by natural aptitude, especially well qualified to work it out. The Mojonnier mechanism, though intricate, is attractive and clean and can be managed under the most simple circumstances.

Students of agriculture will be naturally very much interested in the technicalities of this course, the methods of administration and the ends that it is seeking to accomplish. In general, the course is conducted like any other short course of the University. It is free to residents of Indiana and open to all those women who have completed common school education. The course is administered in Smith Hall, which is thoroughly equipped for special courses. The laboratories are the best and include apparatus for milk testing, cream separation and many intricate devices employed in the dairy industry. All equipment is in first class condition. Upon starting the course, the students are made familiar with all the types of machinery used in caring for milk and cream and their patriotic aptitude is shown and the student is permitted to specialize in a particular phase of the subject. The work is administered by the instructional corps of the regular dairy faculty, assisted by other members of the University. The course has grown out of great demand for trained workers and though the University does not guarantee positions to all those who pass the course, there seems to be no doubt but that successful students will be sure to secure remunerative and pleasant employment.

It is interesting to note that, upon the announcement of the course in the papers, letters began to pour into the office of the Dairy Department from women in the business world who saw the financial possibilities of the work as a profession and who were eager to learn about this new employment. Similar interest has been shown by commercial industries, the large condenseries, creameries and cheese manufactures.[3]

Winter Course, Short Course And Extension

Extension work was, and still is, a large and vital part of the Purdue College of Agriculture. Education of the adult and working public in Indiana was important from the very beginning of Purdue's Agriculture school. Many thousands of students made their way to Purdue for winter coursework and short coursework studies. Winter coursework was normally held in January and February of each year and lasting for about eight weeks. Short coursework normally occurred in a session ranging from a one-day meeting to a week.

As early as 1892, Purdue held winter courses in dairying taught by Professor Charles S. Plumb. These classes were combined with the coursework of livestock husbandry. After three years of combined dairy and livestock classes, dairy husbandry courses were split from the livestock husbandry classwork. In an 1895 brochure advertising the course, Plumb wrote that one or two hours a day would be spent with classwork and several hours each day would be spent with hands-on experiences in the dairy itself. The students would start the day at 8:15 a.m. and finish at about 4:00 p.m. Coursework would terminate at noon on Saturdays. White uniforms and white caps covering the head were required to be worn while working.[4]

In addition to Professor Plumb, several others taught classes, including:

Professor H. A. Huston — The Chemistry of Milk and Its Products
Professor J. C. Arthur — Bacteria and Its Relation to Dairy
Mr. George R. Ives — Milk Testing
Mr. James H. Hilfiker — Butter Making
Mr. Richard Smart — Engines and Boilers

The classes listed here primarily were designed to educate dairymen in dairy product manufacturing while on the farm. All the equipment was designed for farm dairy work as opposed to creamery work. Creamery practice was not taught at Purdue at this time.[5] Professor Plumb also taught a class about breeding and feeding dairy cattle, while Professor A. W. Bitting offered instruction on the sanitation of the stable and diseases of cattle.[6]

Winter Course classes were perhaps as important to Indiana's dairy heritage as the four-year Dairy Manufacturing programs. A listing and discussion of its graduates have been compiled and are attached as an appendix to this manuscript. The 1941 *Debris* denoted the importance of the winter classes, as shown in the following paragraph:

> *The principle purpose of the Winter Short Course sponsored by the Agricultural School is to instruct those ambitious men who are unable to take the regular four year course. . . . yet who desire special training in a particular line of agriculture. Purdue's Winter Course has functioned on campus since 1887. . . . and since that time some 5000 Indiana farmers have benefited.*

The School of Agriculture started these classes in 1887, and Professor Plumb began offering classes specific to dairying in 1892.[7]

The March 1908 issue of the *Purdue Agriculturist* gave a fairly detailed description of the events and happenings of the Short Course for the creamery students. The author of the article is unknown.

> *This is one of the fullest of the short courses. Lectures fill up most of the morning hours every day with the exception of a few hours which, with the hours from one to five on five afternoons, are occupied with practical work in the creamery and laboratories. No pauses are allowed. Something helpful is always ready to be put into any gap.*
>
> *The creamery class has probably the smallest number of men, but this small number gives each one a better chance. The instructors are anxious to see the men learn all that is possible in the eight weeks, and it is really surprising how much is taught. After only three weeks' work, one man stated that he had already learned more about the creamery business than he had expected to learn in the whole course.*
>
> *The cream which is received is not in first class condition, and it has to go through several processes before it will be made into butter which will grade into extras. All of these operations in detail are performed by the creamery students as well as the many incidental operations. Not only does the student learn how to handle the cream and the butter, but also the machinery, belts, pulleys, pumps, plumbing, boiler, etc. He learns the best and practical way of doing everything from washing a cream can to skillfully wrapping the butter prints in parchment and*

> *cartons, from operating the churn gears to soldering a leaky pail. Proficiency with the Babcock test, acid test, and various creamery calculations is employed. A good working knowledge is also acquired of cheese and ice cream. In short the student is fitted to take up the operative creamery work in an intellectual way, and lectures are given which should prepare him for future promotion to the responsibility of managership.*[8]

Dean Skinner, in an article appearing in the December 1925 issue of the *Purdue Agriculturist,* called the eight-week winter coursework in dairy manufacturing, "The most profitable eight weeks of the year." The article talks about how students wanting to become butter makers, ice cream makers, market milk plant operators, dairy plant managers, or superintendents would benefit from the knowledge they gained and the associations they made with new friends and advisors with similar career interests. This article also pointed out that many men already in this field enrolled in this classwork to become more qualified for their employment.[9]

In an article in the *Purdue Agriculturist* in March 1923, the unknown writer laid out the benefits, the costs, and the future for attendees of the eight-week Winter Course in Dairy Manufacturing. The author described how ten of the fifteen men enrolled in the course already had been promised employment. The writer also said that, if the others finish the course satisfactorily, there would be more jobs available than could be filled. In fact, because the university never graduated enough students in the four-year program, this was a problem; so the Winter Course helped meet the needs of the states' employers in dairy manufacturing.

The cost of attending these classes in 1923 was $125. These fees included everything from room and board to laboratory expenses and needed books. Dean Skinner thought of this $125 not as a cost, but as an investment for the future, such as increased earning power over a person's lifetime.[10]

The names and numbers of students enrolled in the four-year program in Dairy Manufacturing has been very difficult to locate. One exception was for the fall classes of 1927. The Market Milk class had four students, Creamery Mechanics had eight students, Cheese Manufacture had three students, and Testing of Dairy Products had six.[11] These numbers can be compared to the eleven Winter Course students who graduated in March 1927 and the nine Winter Course students who graduated in 1928. Student names from the four-year program are also attached as an addendum to this writing.

The Winter Course Bulletin for 1931 denoted the areas of study with detailed descriptions of each specialty. A general description of what would be studied was followed by a description of each individual class that each student would take. The following shows what was written about the Dairy Manufacturing curriculum.

DAIRY MANUFACTURING EIGHT WEEKS
WINTER COURSE
January 19–March 13, 1931

The Eight Weeks Dairy Manufactures Course is a practical course for those who are engaged in a milk, butter, or ice cream plant, or for those who wish to operate or secure employment in a plant handling dairy products.

Those who complete the course have many opportunities available to them, as a number of students have accepted positions as dairy inspectors or as dairy equipment salesmen, and in many other fields which require a knowledge of the practices followed in the handling of dairy products. Students who take this course spend approximately one half of their time in the classroom and one half of the time in the laboratory, where butter, ice cream, and market milk are handled daily on a commercial scale. The Purdue University Creamery is well equipped with continuous and holding types of pasteurizers for milk and cream, churns, homogenizers, ice cream freezers, coolers, ice machines, and other equipment found in up-to-date dairy and creamery plants. All of this equipment is available for students registered in this course.

The subjects offered in the classroom are outlined and taught so as to give the greatest amount of information in the short time of eight weeks. The classroom work is presented in such a way that the students may easily grasp its relation to the practical operations in a commercial plant.

DAIRY MANUFACTURES WINTER COURSE SUBJECTS

351. Butter Making and Plant Management. Factors influencing the manufacture of high-score butter, including cream grading, neutralization, pasteurization, the use of a starter, ripening, churning, salting, working, the control of moisture, overrun, and packing. Dairy plant operation and management. Four lecture and four laboratory hours a week.
Prof. H. W. Gregory and Mr. R. E. Roberts.

351. Ice Cream. Standardization and preparation of the ice cream mix, including mixing, homogenizing, and cooling. Freezing the ice cream mix. Control of the yield. Study of fancy ice cream and ices. Marketing ice cream. Three lecture and four laboratory hours a week.
Prof. K. V. Bryan.

358. Market Milk. Factors influencing the sanitary production, handling, and distribution of milk. Milk plant problems, including pasteurization, cream line, filtering and storage of milk. Two lecture and four laboratory hours a week.
Prof. E. H. Parfitt and Mr. A. V. Moore.

353. Dairy Plant Mechanics. The construction, operation, and repair of mechanical systems of refrigeration and other dairy plant machinery. Two lecture and two laboratory hours a week.
Mr. R. E. Roberts.

357. Testing Milk and Milk Products. The Babcock test for butterfat in milk, cream, buttermilk, and skim milk. Other tests used in standardizing dairy products. Two lecture and two laboratory hours a week.
Mr. A. V. Moore.

359. Dairy Bacteriology. The types and activities of both harmful and beneficial micro-organisms found in milk and milk products. The propagation and use of commercial strains for butter making and cultured buttermilk. Two lecture and four laboratory hours a week.
Mr. A. V. Moore.

300. The Dairy Industry. Extension and investigational work being done in Indiana to improve the dairy industry. One lecture a week. Prof. K. V. Bryan and other members of the Department. [12]

An example of a one-day session was held May 5, 1925. Seventy-five creamery men from throughout the state gathered at Purdue to learn about and discuss manufacturing problems they were experiencing at that time. Mr. Manhart of the Purdue staff discussed the quality of butter being produced in the state. Other speakers from the Purdue Dairy staff were bacteriologist for the department, Mr. E. H. Parfitt,, and the department head, H. W. Gregory. Gregory spoke on the "Factors in Preparing the Factory Churn."[13]

The Dairy Department offered a five-day Short Course for milk plant operators, running from November 11 to November 15, 1940.[14] Twenty-six men enrolled in this class. A near all-star lineup of instructors was assembled. The Purdue Creamery, Purdue Dairy Department staff, and noted dairy owners and operators organized this classwork for Indiana's milk plant operators. Professor V. C. Manhart of the Purdue staff spoke about "Milk Marketing Trends." Professor P. R. Elliker lectured on "Elements of Bacteriological Control," "Milk Borne Diseases," and "Propagation of Cultures and Manufacture of Cultured Buttermilk" and conducted a lab on "Milk Plant Control Tests." Mr. W. H. Brown, an assistant dairy bacteriologist for Purdue, talked about "Platform Inspection of Milk" and "Pasteurization Efficiency" and assisted in a "Milk Quality Testing" laboratory. Dr. B. F. Horrall, a dairy chemist for Purdue, lectured on "Oxidized Flavors and Their Control." Dr. K. V. Bryan, in charge of ice cream instruction, lectured on "Standardizing Milk and Cream."

These preceding speakers were all on the Purdue Dairy Department staff. Other speakers were from the Purdue Veterinary School, private sector dairy owners and operators, and associated dairy industry dignitaries. Perhaps the two most noted and important speakers were Professor Ralph E. Roberts, superintendent of the Purdue Creamery, and Mr. K. C. Boxell, the assistant superintendent of the creamery.

Professor Roberts taught the class on "Market Milk Defects," a laboratory on "Grading and Scoring Market Milk," and "The Processing of Chocolate Milk." Mr. Boxell spoke on "Milk Processing and Distribution Trends," "Vitamin D and Soft Curd Milk," and "The Separation and Processing of Sweet Cream." Mr. Boxell also co-taught the "Milk Quality Testing lab" with W. R. Brown. All sessions for the week were held in Smith Hall, with the exception of the course banquet held in the Purdue Memorial Union.[15]

Classes of this magnitude were not uncommon at Purdue. Another example of this type of gathering was held April 25, 1951. This one-day meeting of The Milk Market Conference was also held in Smith Hall. Subjects under discussion were laboratory control programs for milk plants, milk plant sanitation, the use of glass piping within dairies, concentrated milks, selling cottage cheese, and new state regulations involving dairy plants.

The very next day, meetings of The Ice Cream Institute were held. The latest in ice cream production, storage, and sales trends were discussed. H. F. DePew, a 1917 Purdue graduate and a leading authority on ice cream, was a speaker at both of these conferences.[16]

A slightly different example of Purdue's desire to educate the public took place for many years at the Indiana State Fair. As just one example of that goal, which continues today, the Dairy Department constructed and operated an educational exhibit for dairymen who attended the Indiana State Fair in 1912. This was a two-pronged approach: a section of the display was used to promote proper milk handling on the farm, while the other section of the display was for those interested in dairy manufacturing equipment and techniques for turning milk into products on the farm. Included here were Babcock testers, churns, cream vats, and separators. Throughout the day, demonstrations were provided to the attendees to show the proper procedures for churning, washing, and salting butter. This was followed by demonstrations and lectures on wrapping and packing butter and Babcock testing for butterfat. Apparently, the demonstrations involving the Babcock test were particularly well received because many fairgoers were interested in establishing proper and accurate methods for determining the fat content of milk and cream.[17]

These are but a few of the examples that may have been discussed. Just about every year, or in some cases several times a year, short course programs and extension work were performed for the dairy manufacturing industry.

5

The Years Of H. W. Gregory and Other Dairy Greats

Following the departure of Mills as the head of the Purdue Creamery in 1917, Horace Wilbur Gregory, better known as H. W. Gregory, took over as its superintendent. Like Mills, Gregory also taught classes in dairy manufacturing. Professor Gregory received his bachelor's degree from Oklahoma Agricultural and Mechanical College, today known as Oklahoma State University. He was then employed by South Dakota Agricultural College for four years prior to moving to West Lafayette. In South Dakota, he received his master's degree and was then employed as the college's creamery superintendent where he also taught classes.[1]

Professor Ollie Ezekiel Reed was the Purdue Dairy Department head when Gregory took the job of superintendent of the creamery. Reed began his tenure as department head September 1, 1918. Purdue hired Reed from the Kansas State Agricultural College. Kansas State had previously lured Reed away from Purdue where he had been an instructor of dairying in 1909 and 1910.[2] While at Purdue, he was most noted for increasing the size of the dairy barn on the west side of the campus, adding Guernsey cattle to the Purdue herd, and he formed the very first student dairy judging team.[3]

Shortly after the resignation of Mills, Dairy Department Head Hunziker also left Purdue. The short gap of time between Hunziker and his eventual replacement, Professor Ollie E. Reed, was filled by Dairy Department member Ralph Emerson Caldwell. This gap of permanent leadership for the department spanned the latter half of 1917 and the first part of 1918.

Reed stayed at Purdue only from 1918 to 1921, when Gregory took over the responsibilities of the entire Dairy Department. The time of transition from Reed to Gregory as Dairy Department head is full of unknowns left by the various historical documents of the era. The first known notice of Reed's resignation comes on April 7, 1920, as found in that issue of *The Jersey Bulletin and Dairy World.*[4] According to the Purdue Board of Trustees meeting, he turned in his resignation on April 8, the day after it was released in *The Jersey Bulletin and Dairy World.* It was officially announced at the April 14 Board of Trustees meeting with no date set for his departure.[5] The board's June 8 minutes announced that the Dairy Department position was open.[6] According to the *Purdue Agriculturist*, Reed left his post August 1.[7] H. W. Gregory was then appointed by the Board of Trustees to fill the vacancy on October 6, 1920. The Board of Trustees also announced that Gregory would receive an annual salary of $3,750.[8]

Equally murky is the transition from Gregory as the superintendent of the creamery to the appointment of the succeeding manager. C. O. Tuttle is believed to have taken over management of the creamery some time during 1921.[9] Tuttle was on the job when classes started in the fall of 1921. What about the gap in time between October 1920 and the time Tuttle took over? There is only one reference in the Board of Trustee minutes (June 1922) of Tuttle's employment at Purdue. Could there have been an interim manager during that time gap?

Two references appear in the *Purdue Agriculturist* that indicates what might have happened. The following from the May 1924 issue offers us one clue:

> *H. W. Fleisher, formerly of Schlosser Bros Creamery of Fort Wayne, and later superintendent of the Purdue Creamery is now manager of the Lafayette Ice Cream Company.*[10]

A January 1929 issue of the same publication gives us the second clue. This statement says similar things as the first, but the order of employment was changed:

> *H. W. Fleisher of the class of 1918 was connected with the Creamery License Division at Purdue following his graduation. He became Superintendent of Schlosser Creamery Company at Fort Wayne. Manager of the Lafayette Ice Cream Company at Lafayette, Indiana, he then became superintendent of the Purdue Creamery at Purdue and is now located with the Freeman Dairy Company, at Flint, Michigan. He is married.*[11]

Fleisher resigned his inspectorship job with the Creamery License Division between March 31, 1920, and March 31, 1921. The exact date is not known. We do know that Fleisher was married in June 1921 to a local Lafayette girl, fitting the time frame.[12] We also know that he was a Kendallville, Indiana, native, and his given name is Hurbert W. Fleisher.[13]

There is stronger evidence to suggest he was more than just an interim superintendent or manager. It is given by the *Purdue University Annual Catalogue for 1920–1921*. He is listed as a person teaching several dairy and creamery courses for that academic year. His name is not listed in either the previous year or the following year of the Purdue Annual Catalogue. Under the Agricultural Experiment Station Staff in the same publication, Fleisher is listed as "Assistant in Dairy Manufactures." The word assistant in this reference was in regard to rank of professorship.[14]

In the phonebook of students at Purdue during the 1920–1921 school year, entitled "Directory of Purdue Students 1921," Fleisher's name can also be found, not as a student but as a staff member. Not only did this publication have the student names and numbers, but also staff members, buildings, and department phone numbers are given. In these pages "Creamery" is listed with H. W. Fleisher as the person to call in regard to the creamery. This was the only year with Fleisher's name shown under this heading.[15] In 1920, the name H. W. Gregory is found and in 1922 C. O. Tuttle's name is listed.[16,17] We know that Gregory and Tuttle were both Creamery superintendents during the years they were listed, so it is easy to conclude that Fleisher would have been as well.

One other piece of information linking him to Purdue, but not to the question of being superintendent, was that he was employed to teach Winter Course students for eight weeks starting January 24, 1924. His salary was $25 a week for eight weeks.[18] Additionally, in a letter dated September 28, 1925, Professor Gregory wrote to Dean Skinner that Fleisher had turned in an application for graduate work.[19] This was the final known connection between Fleisher and Purdue University. Whether Fleisher filled in as interim superintendent may truly never be known. No reference to the title of Creamery Head or Superintendent has been found.

Very little is also known about the operation of the creamery during Mr. Tuttle's tenure as superintendent. Because he was a member of the 1913 Purdue graduating class, the 1913 *Debris* reveals some details about both his college life and personal life.

> *Tut is a quiet, genial, and unassuming person. He has the rare ability to attend to one man's affairs, and consequently he has many friends and acquaintances. He has never formed the fussing habit but in the last year has shown a marked tendency that way. By his conscientious work he has made a good record in the classroom. With the benefit of his practical experience along the line of his given work, he will no doubt make a successful farmer.*

The publication also reveals that the Warsaw, Indiana, native's given name was Carl Ogden Tuttle. He was a member of the Hoof and Horn Club and Alpha Zeta.[20]

In regard to the reference to Creamery License, milk and cream purchased within Indiana from dairy farmers must be accurately weighed, sampled, and tested for butterfat. This is to ensure fair and honest payment from processors to the dairy farmer. Indiana code, beginning in 1914, assigned this enforcement to the Creamery License Division under the direction of Purdue University School of Agriculture. Other references to Creamery License will appear later as well.

On June 1, 1922, Tuttle was given a $100 raise, increasing his yearly salary to $1,600.[21] Mr. Tuttle resigned as creamery manager November 15, 1924.[22]

When C. O. Tuttle resigned as superintendent of the creamery he had been employed in this position for four years. He had been employed prior to his Purdue stint from 1917 to 1921 by Schlosser Bros. as the head of all the company's Indiana creameries. When Tuttle left Purdue, he joined the Detroit Creamery Corporation. This move was probably viewed by Tuttle as a promotion or advancement in position, as his new employer handled over half of the business of Detroit in both butter and ice cream. The Detroit Creamery Corporation produced more than five million pounds of butter a year.[23] In comparison, Purdue was producing approximately 300,000 pounds per year.

Purdue promoted Professor Gregory from Creamery superintendent to Dairy Husbandry Department head. His tenure as Purdue Dairy Department head lasted for forty years, running from August 1920 to 1960. At the time, Gregory was perhaps the perfect man as department head because he blended his knowledge of dairy manufacturing with his on-the-farm production experience. During his tenure as department head, Purdue moved to the forefront in dairy research, education, production, and extension.[24]

H. W. Gregory (left) with Governor George Craig (seated) at an Indiana Statehouse ceremony.
(Courtesy of John Cleland)

The demand for information concerning dairy product processing was growing rapidly during this time period. This growth was evidenced by the 2,500 people who toured Smith Hall and the Purdue Creamery in 1921.[25] Purdue was advertising as early as 1911, when the creamery was in the Agricultural Hall basement, that visitors were welcome during business hours.[26] (The creamery policy had apparently been to welcome visitors.) The public and the people of the industry were seeking information, and this demand was growing rapidly. The dairy processing industry was becoming more scientific, efficient, and sanitary. The industry was changing drastically with new product development and increased mechanization.

In the 1920s, Gregory's leadership attempted to meet these demands by enhancing the creamery through several equipment upgrades and changes. Purdue wanted to stay ahead in this game, and beginning in January 1921, the creamery installed new vats for the manufacturing of buttermilk. These vats were glass-lined and enameled to help prevent corrosion from the acids found in buttermilk.[27] Purdue's Board of Trustees quickly followed this with $5,000 allocated in April 1922 for additional upgrades to the equipment.[28]

In the early part of 1923, the creamery installed a most up-to-date piece of equipment for the sterilization of glass milk bottles. This sterilizer guaranteed the cleanliness of these bottles prior to filling and delivery to the customer. The operator would enclose the containers within the machine and expose them for thirty minutes to steam at 212 degrees Fahrenheit. It had a capacity of up to 360 half-pint bottles at one time. Very few dairy plants owned one of these sterilizers at this point in time.[29]

A new 40-quart ice cream freezer was installed in the creamery in December 1923. This new freezer was noted for the uniformity of its frozen product. The freezer was the latest and most up-to-date model manufactured by the H. N. Miller Industries Company of Toledo, Ohio. It was equipped with a 3 HP motor, a self-oiling chain guard, gears of forged steel, and a fruit funnel. This funnel held and delivered different fruits that were added to help flavor the ice cream.[30]

December 16, 1923, the Purdue Creamery reached an agreement with H. A. Hovey and Co. to ship 900 pounds of butter a week to the

New England states. Most of these shipments were sent to Boston, Massachusetts. The creamery received two cents a pound over the local wholesale butter market.[31]

November 15, 1924, the same day Tuttle's resignation took effect, Ralph E. Roberts was appointed head of the Purdue University Creamery. Mr. Roberts was a Lafayette, Indiana, native who attended Ford School and graduated from Jefferson High School. He had been president of the Purdue Dairy Club, manager of the Dairy Club Demonstration Night, and a Dairy Products Judging Team member. He was also active in the class of 1921 as the art editor for the *Debris* and a member of Alpha Zeta, which is still a School of Agriculture scholastic honorary fraternity.[32] Roberts was a 1921 Purdue dairy manufacturing graduate who had been employed by Purdue to teach dairy manufacturing and dairy production in a three-year program to special students sent to Purdue by the United States Veterans Bureau.[33] These federal students were in a rehabilitation program designed for wounded and disabled American service veterans of World War I. One hundred twenty-five men were enrolled in this program offered by the School of Agriculture with eighteen specializing in dairy manufacturing and seventeen specializing in on-the-farm dairy production. Ten of the thirty-five completed coursework in both areas. Thirty-four men outside the area of dairying, but taking other classes offered to the federal students, elected to take at least one dairy class. A total of twenty different dairy classes, both in manufacturing and production, were offered in thirty-seven sections during the three years the program was in effect.

Ralph Roberts.
(Courtesy of Purdue University Archives and Special Collections, Purdue University Libraries)

Twenty-seven men who were enrolled in this U.S. Veterans Bureau course later passed the examination for an Indiana Cream Testers License administered by the Creamery License Division. This allowed these men to perform the Babcock test to determine how much butterfat is found in milk and cream. This three-year program ended in 1923. It was reported in November 1924 that nearly all of these federal students had found employment on their own, or that based on their strong recommendation, Mr. Roberts and Professor Gregory had already secured employment for them.[34] After this program finished, Roberts continued with the Purdue Dairy Department as an instructor in dairying.

In 1923, experimentations on improving butter quality were still being conducted. One example was the use of a water softener to produce soft water for the washing of butter. The Purdue staff and creamery were trying to determine whether soft water improved the keeping qualities of the butter against that of the traditional hard water.

Another experiment conducted on butter focused on the relationship between the effect of bacteria and the length of time it took for butter to become rancid. Purdue used sixty different types of bacteria that were isolated from milk to test this relationship. A third experiment using sweet cream in butter making was also done.

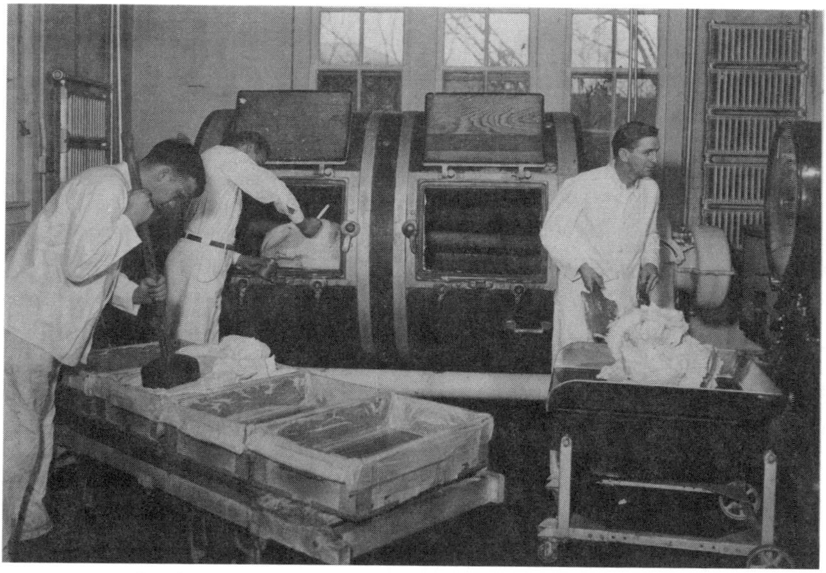

Purdue students making, weighing, and packing butter.
(Courtesy of Purdue University Archives and Special Collections, Purdue University Libraries)

Other investigations were taking place as well, including the adulteration of butterfat and finding methods for the detection of the materials causing this adulteration. Purdue was also doing experimental work with bacterial enzymes on milk proteins.[35] Conclusions were never finalized on these experiments, so the results were never published. One other test being conducted involved a bottle sterilizer. Purdue was attempting to improve the quality and longevity of milk in bottles for their home delivery routes because spoilage and product quality had always been problems in dairy processing.[36]

In a letter to Dean Skinner dated December 31, 1924, Department Head Gregory discussed needed equipment and new educational and experimental endeavors for the Dairy Department. He was seeking new equipment for milk condensing as Purdue was almost totally devoid of any experimentation or educational activities in this area. Equipment for student instruction in butter making, ice cream making and market milk laboratories was worn out and woefully out of date. Gregory also thought it was quite embarrassing for the university to have dairy plant personnel show up for short course instruction and find equipment that was far less modern and much less efficient than they had at their own plants throughout the state. Gregory asked for $20,000 dollars for these upgrades.[37]

New equipment for the creamery arrived in late 1925 and was ready to be installed. Of special note were a vertical coil Jensen Milk vat and two holding vats manufactured by the Creamery Package Company. These vats had inside surfaces consisting of blue glass laid on top of steel, which made these vats more sanitary, gave them increased durability, and extended working life. Also delivered were a Jensen fore-warmer, Simplex churns, and two Pfauder pasteurizers. In addition, a Creamery Package rotary can washer and a Ft. Wayne continuous milk and cream pasteurizing device were installed. All of this new equipment was powered by electrical motors. This new and improved equipment added great value to the educational agenda of the creamery.[38]

Purdue also researched improvements in the manufacturing of ice cream. As popular as Purdue ice cream was in the later years in the operation of the creamery, sales were not all that great during 1929, with only $4,194.26 in sales. This was small compared to over $120,000 in butter sales, $18,000 in sweet cream sold, and $13,000 of milk sold.[39] New tests sought the causes of sandiness in ice cream texture.[40] During 1929 and 1930 studies were conducted at Purdue on a variety of areas in the production of ice cream. These areas included the pasteurizing temperature of the ice cream mix, the firmness at which to draw the ice cream, the storage temperatures of ice cream, and the effects of temperature changes in the transportation and delivery of ice cream.[41]

Students making ice cream.
(Courtesy of Purdue University Archives and Special Collections, Purdue University Libraries)

In a letter dated March 31, 1926, President Elliott asked if the Dairy Department and Purdue Creamery staff would be able to increase the volume of business because he wanted the creamery and the dairy farm to operate on a self-supporting basis.[42] Financial records for the years prior to Elliott's letter have not been located, but records for 1926 and later were found. They show a $3,290 profit in 1926[43] and a $299 gain in 1927.[44] In 1928 a loss of $2,243 occurred[45] with another loss of $1,126 for 1929.[46] For the third year in a row, a loss of $2,140 took place in 1930.[47] After this point, the creamery profits turned around, and the only years in which losses happened were in 1935 and in 1969, which was the final year of creamery operation.

During the latter half of the 1920s, President Elliott became disturbed by the cost of operations and effectiveness of both the creamery and the Purdue Dairy Farm. Both of these were costing more money to keep in operation than he apparently desired to spend. President Elliott also wanted to know how important it was to keep creamery machinery and equipment updated on a regular basis in regard to student learning and experience.[48] On September 26, 1926, Elliott strongly encouraged major changes in the management and leadership of the Purdue Creamery, the dairy farm, and the dairy herd to make them more profitable. Later in 1926, Professor L. H. Fairchild felt this pressure as the person in charge of the dairy farm and resigned this position and left Purdue.[49] Gregory as Dairy Husbandry Department head and Roberts as Creamery superintendent both survived and kept their jobs.

In October of that year, $3,600 was allocated to purchase equipment and paint the creamery interior. The entrance to the creamery was also remodeled.[50] In 1926, $25,000 more was spent to replace obsolete machinery. This was an extremely large sum of money for the time and was the largest amount yet spent on the creamery.[51] This amount of money was roughly equivalent to $334,000 in 2014.

Gregory represented Purdue at the prestigious World Dairy Congress held in 1928 from October 2 through October 12. The meetings were held in Washington, D.C., October 2 and October 3 and in Philadelphia October 4, and the final week of meetings were held in Syracuse, New York. Gregory spoke to the leaders of the world's dairy industry about creamery and milk testers' licensing laws in Indiana. Indiana was well known for its regulatory management of creamery and milk testers' licensing. The enforcement of these laws was administered by Purdue with the Dairy Department head serving as chairman of the State Creamery License Board. This law was enacted by the Indiana State Legislature in 1914.

Abstracts of the presentations and papers presented to the World Dairy Congress were translated into English, French, German, and Spanish so delegates from around the world could receive and understand the latest in dairy information, including Professor Gregory's lecture.[52]

Gregory's skills as a judge of dairy products were also occasionally on display. He judged many contests around the state and Midwest.

He instructed students during classwork in this skill. It was considered an honor and a reward to be selected as an official judge for the Collegiate Students International Contest in Judging Dairy Products. Gregory served as the judge for butter on numerous occasions for this contest.[53]

Gregory's reputation and ability to judge dairy products still wasn't an automatic ticket to take off anytime he wanted. Dean Skinner still had to approve such travels. A letter from Skinner to Gregory on August 29, 1931, showed his desire to keep Gregory close to home. Within the letter Skinner "reluctantly approved" Gregory's request to be gone September 11 and 12 to judge at the Kentucky State Fair. With the fall semester starting and "other things happening," Skinner was concerned about Gregory's absence for two days.[54]

It is easy to forget that one of the main goals of having a creamery at Purdue University was to educate students in the safe and economical production of dairy products. Over the course of time in which Dairy Manufacturing was taught, only slight changes were made in the type and arrangement of classes offered to students. For instance, Dairy Bacteriology was taught for the duration of the program, but new knowledge was continually added and given as part of the instructional process. Creamery Mechanics was also offered throughout the years, but equipment was often modernized and updated; therefore the instruction was also modernized. Information on Dairy Husbandry and Dairy Manufacturing from the 1930 Purdue University Bulletin is shown below as an example of what was taught.

The 1930 *University Catalogue* described the four-year plan of study for the School of Agriculture as follows:

> *The four-year plan of study offers opportunity for a broad, scientific education and for a certain amount of specialization in some chosen field of agriculture. A large number of courses of instruction in technical agriculture are given by the Departments of Agricultural Engineering, Agronomy, Animal Husbandry, Dairy Husbandry, Entomology, Farm Management and Rural Economics, Forestry, Horticulture, and Poultry. Agriculture students receive instruction in the sciences of botany, zoology, chemistry, and physics, which are fundamental to agriculture work, and in English, history, literature, education, economics, and mathematics.*

> *Certain courses are required of all students in each of the four years: others are elective and may be chosen upon approval of the Dean and head of Department. Freshmen are required to attend weekly assemblies. Juniors and seniors may be required to take inspection trips in connection with certain plans of study.*
>
> *By proper choice of electives students may fit themselves for practical work as farm operators or managers, or for scientific work in fields of agronomy, agricultural chemistry, animal husbandry, agriculture education, dairy production, dairy manufacturing, forestry, horticulture, entomology, floriculture, vegetable gardening, farm management and agricultural economics, and poultry.*
>
> *While the agricultural courses are designed to train students for farm and life in the country, they may at the same time fit men for a variety of very excellent positions in educational, scientific, and commercial work in connection with institutions and industries closely allied with agriculture.*[55]

The same 1930 *University Catalogue* describes the Dairy Husbandry and Dairy Manufacturing options as follows:

> *The dairy manufacturing option is suggested for students who wish to specialize and prepare for positions requiring training in the manufacture of butter, ice cream, cheese, market milk, and condensed milk. This option prepares the students for positions in teaching and extension, and for numerous government and commercial positions which require dairy manufacturing training.*[56]

The 1930 Bulletin also describes courses taught for Dairy Husbandry and Dairy Manufacturing. As written earlier, Dairy Manufacturing was part of the Dairy Husbandry Department; thus many of the classes overlapped. A student in one discipline would often take coursework in the other discipline of study. The numbers next to the class names below may not always be consecutive as only Dairy Manufacturing courses are listed with Dairy Production coursed omitted.

DAIRY HUSBANDRY

1. ***Elements of Dairy Production** — Survey of the industry; the fundamentals of raising, managing, selecting, breeding dairy cattle; care of milk and cream on the farm; milk testing; dairy cattle judging.*
   Professor Bryan and Mr. Moore.

2. ***Creamery Butter Making and Factory Management** — The theory and practice of creamery butter making. Laboratory work in preparation of cream, operation of churn, preparing butter for market.*
   Mr. Roberts.

3. ***Testing of Dairy Products** — Determination of fat, salt, moisture, acidity, casein, total solids, and solids not fat in milk and its products; standardization of milk, cream, alkaline solutions and acid; detection of adulterants and preservatives; quality tests of milk. Laboratory work.*
   Professor Bryan.

5. ***Ice Cream and Condensed Milk** — The manufacturing and marketing of ice cream and frozen products and of condensed and powdered milk. Laboratory work. At least one inspection trip.*
   Professor Bryan.

8. ***Milk production** — History and distribution of dairy breeds in the United States; selecting and developing the dairy herd; rations; the value of home grown feeds; the arrangement of the dairy farm.*
   Professor Hilton.

10. ***Creamery Mechanics** — Factory construction; the various machines used in creameries. Laboratory work used in the examination, setting, and operation of these various types of machines.*
    Messrs. Roberts and Moore.

***11. Cheese Manufacture** — The manufacture and marketing of various fancy and soft cheeses. Laboratory work in the manufacture, care, and storage of cheese; practice in scoring and judging.*
Professor Bryan; Mr. Moore.

***13. Dairy Survey** — For students specializing in dairy manufacturing. The costs of producing milk; breeds of dairy cattle; the pure-bred bull; co-operative organization; the construction of dairy farm buildings. Laboratory work in judging dairy stock and designing dairy farm buildings*
Professor Wilbur.

***15-16. History and Breeding of Dairy Cattle** — Origin, history, and development of dairy breeds; their distribution in Europe and America; co-operative cow testing and breeding associations; official and advanced registry testing. Laboratory work in judging dairy cattle and in tracing pedigrees.*
Professor Wilbur.

***17. Elements of Dairy Manufacturing** — Composition and methods of manufacturing dairy products; care of milk and milk products; separators; cream and buttermilk testing; factors which determine quality of dairy products.*
Professor Bryan; Mr. Moore.

FOR UNDERGRADUATE AND GRADUATE STUDENTS

***104. Market Milk** — Training in the production of market milk, and for work in city milk plants, public health and inspection service. Laboratory work in scoring and inspecting barns and milk plants, and in testing milk and milk products.*
Professor Parfitt.

***107. Care and Management of Dairy Cattle** — Development of dairy herds; care, management, and methods of feeding dairy cattle; dairy farm buildings. Laboratory work in identification and preparation of dairy feeds; dairy barn construction plans, and dairy herd management problems.*
Professor Hilton.

***109. Dairy Bacteriology** — The significance of bacteria in milk and milk products, and the laws relating thereto; equipment and methods used in commercial and public health laboratories. Laboratory work in preventing bacterial contamination of milk and milk products*
Professor Parfitt.

***112. Dairy Chemistry** — The chemistry of milk and its products; feeding and its effects upon the composition of milk: chemical changes involved in butter and cheese manufacture, and factors affecting their composition. Treated from the standpoint of biological chemistry.*
Mr. Brewer.

***114. Advanced Manufacturing** — Technical problems related to the manufacture of dairy products; methods of experimentation. Laboratory work along a chosen line of dairy manufacturing, complete outlines and reports of all assignments.*
Professors Gregory and Parfitt.

***118. Advanced Dairy Cattle Feeding** — Literature relating to feeding of dairy cattle; physiological study of the dairy cow; value of individual feeds used in compounding dairy rations; grinding and mixing of feeds; problems in feeding dairy cattle.*
Professor Wilbur.

PRIMARILY FOR GRADUATE STUDENTS

***201. Seminar** — Formal report and discussions of selected topics to dairying.*
Professor Gregory and Staff.

***203. Topical Research Problems in Dairy Production** — Topics of interest to the students and training in investigation and research in dairy production.*
Professors Hilton and Wilbur.

204. Topical Research Problems in Dairy Manufacturing — *Topics of interest to the students and training in investigation and research in dairy manufacturing. Professor Parfitt.*[57]

Prior to his retirement in 1939, Dean Skinner sent the university his budget request for 1939 to 1941. In this budget, he asked for $60,000 to be used for remodeling, updating, and additions he thought needed to be made to Smith Hall. In his written request, he provided rationale for the urgently needed extra space and facilities.

Needs for Additional Laboratory Space

Remodeling and adding an addition to Smith Hall would permit an increase in space for our bacteriology laboratory which is so small at the present time that it is necessary to divide classes into two sections, which doubles the time of the instructor in order to take care of the students.

We should have separate laboratories for graduate students, research, and undergraduates. Such an arrangement would increase the efficiency of our instructional and research work. Entirely too much time is now devoted to rearranging and setting out equipment in our small laboratory for these groups.

We have at the present time no laboratory for condensed milk or curing room for the manufacture of cheese. There are only two other states making more cheese than Indiana and requests for instruction and experimental work in this subject are increasing.

Needs for Additional Office Space

Our office space is very limited in Smith Hall. We have from two to four people in each office. This crowded arrangement is not conducive to efficient work by our research workers and teaching staff.[58]

In August 1939 and over the next several months, Professor Gregory contemplated expanding Smith Hall and the creamery. He wrote several letters to the new Agriculture dean, H. J. Reed, who had taken over from Dean Skinner. Gregory also wrote to the Purdue University controller, R. B. Stewart. Gregory was concerned about the lack of storage space and overcrowding of students, and he wanted up-to-date equipment in Smith Hall. Gregory suggested that "new and convenient storage be provided." He also suggested that a new elevator or a new stairway up to the attic be constructed for easier access to the storage space for supplies. Gregory also reminded Dean Reed and the controller that the Dairy Department also had five or six trucks that needed shelter and should be taken into account when building storage space.

It is evident in his first letter dated August 4, 1939, that Gregory's favorite solution to this problem was to add onto Smith Hall: "The best solution, as far as the Dairy Department is concerned would be to build an addition to Smith Hall and to provide needed laboratory and classrooms and offices and use the present cheese laboratory for storage space and the attic in a limited way as suggested earlier."[59]

Mr. Henry B. Abbett, purchasing agent for the university, became involved in a portion of this discussion when he ordered an inspection of the Smith Hall elevator by the Otis Elevator Company.[60] It was discovered during the inspection that the elevator needed to be replaced as quickly as possible. Otis Elevator representative C. T. Spriggs wrote in a letter to Mr. Abbett, "for there is a possibility of a serious accident." Spriggs later wrote, "It is our recommendation that this elevator be replaced with a modern type traction machine with suitable capacity and speed and with all parts of the equipment complying with the existing State Elevator Code."[61]

Controller Stewart jumped into the discussion by saying, "I personally have been critical of the rather patching up method we have followed for some time and because it seems to me that we have spent a lot of money hither and yon without knowing definitely toward what we were building." He went on to object to having a storeroom in the attic and the use of the elevator as "most inefficient." He also called the elevator "a menace to life."[62] On October 8, 1939, Dean Reed responded to Mr. Stewart with a letter of his own. In this letter, Reed mentioned that he was going to secure an estimate for replacement of the elevator and reminded Stewart that the elevator was a freight elevator and not for carrying human passengers. Reed also told Stewart that he believed adding an extension to Smith Hall would solve the overcrowding problem. He also felt it would increase the attic space and consequently, storage space. This additional storage space would make "the elevator essential."[63]

Three days later Gregory asked Reed if he found the report of August 4, 1939, acceptable. If Reed did find the report acceptable, Gregory suggested that they hire an architect to draw up plans to alleviate the problems of lack of space and storage.[64]

Less than two weeks later, Gregory sent another letter to Reed with drawings showing his suggested additions to Smith Hall along with the location. These drawings were apparently done directly by Gregory, as he also suggested that "they be gone over by an architect."[65] These plans have not been located. Copies of these plans were sent by Mr. Reed to Mr. Stewart and to President Elliott. In this letter, Reed sent two plans to Stewart: one with the elevator discontinued and the second with the design the Dairy Department personnel were proposing. Reed also wrote about how the training in cheese manufacturing had been severely handicapped for "our Indiana boys" as facilities and coursework were lacking in cheese manufacturing. Cheese production was rapidly becoming an important aspect of Indiana agriculture. He goes on to write:

> *It will be necessary sooner or later to enlarge Smith Hall. The Department does not have adequate room at the present time and work of the school of Agriculture increases and the attendance of our short courses becomes larger; we must make provisions for adequate training.*[66]

Mr. Stewart responded in a letter dated January 2, 1940, and again showed his opposition to the elevator by suggesting funds for an elevator instead be used for construction of additional storage someplace else. Mr. Stewart wanted action to take place rather quickly as he wanted to present possible plans to President Elliott in the near future. His plan was for Elliott to seek action from the Purdue Board of Trustees at the next board meeting later in the month.

An interesting item is handwritten at the bottom of this letter: the word "basement" written in red pencil and underlined. Who wrote this comment remains a mystery. Was it Stewart while thinking about another option for storage space? Was it Reed, the recipient of the original letter, pondering other options? Or was it Gregory, who also received a copy of the letter?

Dean Reed's response to Stewart in a letter dated January 8, 1940, does little to solve this mystery:

> *My Dear Mr. Stewart:*
>
> *I discussed with Prof. H. W. Gregory your letter of January 2, regarding storage space in the proposed addition to Smith Hall.*
> *Professor Gregory says that a basement under the proposed addition would give ample space to take care of the material which could not be stored in the room indicated for storage in the plan you have. It was thought that the heavy storage, such as salt, etc., could be kept in the room now used as a miscellaneous laboratory for cheese, etc. This room has an outside entrance for receiving, and it is near the departments using this heavy material. The lighter storage could go in the basement, and no elevator would be required.*
> *I realize that at the present Smith Hall does not have a basement, but I see no reason why such a room could not be constructed reasonably. The same thing occurred in the addition to the Horticultural Building.*
> *If there are further questions, I would be glad to arrange a conference in your office with Professor Gregory.*
>
> *Very truly yours,*
> *H. J. Reed*

This is the last known correspondence between these two concerning Smith Hall.[67] A letter from Gregory to Reed dated February 16, 1940, states that Gregory sent in a requisition for installation of a stairway and elevator in Smith Hall, even after all the previous discussion of other options. This letter stated that the Purdue Board of Trustees and Mr. Stewart had approved these plans, but that Purdue Creamery funds were to be used. Gregory also said that the creamery funds were adequate to take care of this project and that he wanted the construction to be completed as quickly as possible because "the present elevator is dangerous."[68]

President Elliott, in January 1940, requested that the university creamery trademark be copyrighted.[69] In April of the same year, the university announced that all dairy products served on campus were produced in the Purdue Creamery. The Board of Trustees' policy was to give students access to all the high-quality dairy products they wanted. The university also announced that no butter substitutes were purchased for use on campus.[70] Also reported in the board meeting were the income and expenses of all the self-sufficient operations on campus. Those operations were listed as the residence halls, the Memorial Union, the University Creamery, and intercollegiate athletics. No part of the income of $1,176,223.91 from these sources was from the state of Indiana. The expenses were $1,155,557.89, meaning the total profit was $20,666.02 from these four enterprises.[71]

The leadership of Purdue University, and particularly Professor Gregory, was noted in a *Purdue Exponent* article concerning butter research and the improved quality of butter produced throughout Indiana. From its inception in the 1890s, Purdue continually strove for higher quality butter production. The April 4, 1941, *Exponent* article mentioned a gathering of one hundred creamery men at a one-day meeting on campus where Gregory reported on the progress made in butter research. At this meeting, Gregory also informed the group that ninety percent of the Indiana butter was being produced by fifty creameries.[72]

The production and quality of butter continued to be important not only to Purdue University but also to dairy and creamery men throughout Indiana. It was apparently important to the general population on campus as well. Purdue student J. R. Hobson even went so far as to write poetry about butter. His poem was simply entitled "Butter" and appeared in the May 1940 *Scrivener.*

BUTTER

By J. R. Hobson

The skim milk settles to the bottom.
That is its nature.
It could not do so otherwise.
The cream rises to the top.
That is its nature.
Put it into a churn, it is beaten-
Because it is good,
It is tormented.

Some of the cream is discouraged.
It goes back
To live the untroubled life of milk
(Cynical milk!)
But some of the cream has courage.
It perseveres.
It triumphs.
It
Is Butter[73]

6
The Great Depression, Forecast For The Future

Economic times became tough in Indiana and the entire country starting in October 1929. As the Great Depression hit, businesses and individuals began to struggle to meet financial crises. These financial failures did not seem to affect the university's creamery to a great extent.

Despite the difficult times brought on by the depression of the 1930s, the creamery did well, averaging more than $3,000 in profit annually while building up a substantial reserve. This reserve money was used to pay for needed repairs, replace equipment, and avoid the use of taxpayer money for the creamery. In 1931, a profit of $4,307 was made, the largest for many years.[1] Purdue was ramping up and expanding creamery production during the 1930s. This was evidenced during the 1932–1933 school year by the processing of more than 121,000 pounds of butter, 20,000 pounds of cheese, 6,300 gallons of sweet cream, almost 28,000 gallons of buttermilk, 60,500 gallons of milk, and more than 7,600 gallons of the famous Purdue ice cream. During this same twelve-month period, Purdue purchased 138,055 pounds of butterfat or cream to be used in product manufacturing during the nine months students were on campus.[2]

This success of Purdue's dairy in the early part of the depression led thirty-one local dairy and business owners to protest in a letter to Purdue about the university marketing dairy products to the general public in competition with taxpaying dairies and allied enterprises. In that letter, dated October 6, 1930, they also wrote that Purdue was using deceptive practices by buying milk and cream from local farmers and marketing the finished product under the Purdue label. They suggested that this led the purchasing public to believe all the milk and cream was produced by the Purdue dairy herd. They suggested that the products produced by the creamery be sent to the Indiana State Soldier's Home, Indiana State Children's Home, other state tax-supported institutions, or possibly to other markets not competing with Lafayette area dairy dealers.[3]

Dean Skinner and Professor Gregory formulated arguments to rebuke the main ideas of the letter and presented them to President Elliott. Armed with the advice of Skinner and Gregory, Elliott responded in a letter of his own on December 19, 1931, addressed to the thirty-one complainants. He wrote that no taxpayer money was used in the operation of the creamery and that the creamery was self-supporting.[4] He also stated that students were being taught real dairy manufacturing techniques and good business methods using large quantities of product. He wrote that there was no deception taking place because the purchasing public knew the Purdue dairy herd was small and did not produce enough milk to supply all the products demanded. In regard to supplying other state-supported entities, Elliott pointed out that these institutions also had dairy herds that met their own needs for dairy products. In closing this letter, Elliot wrote about how the creamery had supplied assistance to these protesting private dairies of Tippecanoe County for many years and was willing to do so in the future. He also believed the creamery was treating all parties engaged in dairy activities with fairness.[5] In June 1934, University Controller R. B. Stewart suggested that the Purdue Memorial Union, the Purdue Creamery, the Horticultural Department, and Animal Husbandry join the Lafayette Chamber of Commerce to better represent themselves when matters that pertained to their own activities came up for discussion.[6] This idea was not only to help represent themselves but also to suggest to the local businesses and entire community that Purdue wanted to be a cooperative business partner within the Tippecanoe County area.

It was during this time that Earl Butz, later to become both the Purdue Dean of Agriculture and the U.S. Secretary of Agriculture, remembered buying ice cream cones as a student from the salesroom in Smith Hall. Ice cream cones were only a nickel while he was enrolled from 1928–1932. Butz hinted that no matter how tough times were, he could always scrape together five cents for ice cream. In 1932, Earl Butz graduated from Purdue in Agricultural Economics. He recalled these memories in a 2001 interview with Jack Albright.[7] This five-cent price stayed in effect until 1968 when it doubled to a dime.[8] Even with the doubling of price, Merle Cunningham, professor emeritus in Animal Science said, “You never walked away hungry after having a double dip. You always got more than your money’s worth with chocolate always being my favorite.”[9]

It was during the Great Depression that one of Purdue’s most noted dairy graduates finished his undergraduate degree in 1933. Wendell S. Arbuckle, of Scottsburg, later in life became known as “Mr. Ice Cream.” Arbuckle traveled to Columbia, Missouri, and earned his master’s degree in 1937 and his Ph.D. in 1940, also from the University of Missouri. He became known around the country and the world as the leading authority on ice cream.[10]

After being awarded his doctorate, Arbuckle joined the faculty at what is now Texas A&M University. His second faculty position was with North Carolina State University. He finally settled at College Park, Maryland, with the University of Maryland. From 1949 to 1972, he conducted research on ice cream. His experimentation investigated its chemistry and crystalline structure and led to the development of many new flavors. Some examples are sweet potato, rhubarb, carrot, bubble gum, holly berry, cantaloupe, corn, and even grass. It was noted that grass flavored ice cream never became a commercial success.

Dr. Arbuckle authored or coauthored more than 200 scientific articles on ice cream and dairy products. He also wrote several books, including *Ice Cream*, his best-known work, which became the standard for the industry.[11] Even Jerry Greenspan of Ben and Jerry’s Ice Cream used a copy of this book to help lay the foundation for its well-known ice cream business.[12] Upon retirement from the University of Maryland in 1972, Arbuckle and his wife started Arbuckle Inc., a private ice cream consulting firm. Dr. Arbuckle passed away in 1987.[13]

With the difficulties brought to the average person during the depression, the university and the creamery were sensitive to these financial hardships of citizens within the local community. In response to a request for resources from the Lafayette Salvation Army, Controller Stewart and President Elliott suggested that excess buttermilk be given to the Salvation Army to help feed the hungry of Tippecanoe County. The Salvation Army requested twenty to thirty gallons per week, but the creamery was able to provide only ten gallons per week.[14]

While the Purdue Creamery may have been responsive to the needs of the local citizenry during these trying times, it was less sensitive to a financial problem presented by one of its own. One of its West Lafayette home delivery route drivers was struggling to meet his payments to the creamery. This route of 200 homes was being serviced by a private distributer instead of a Purdue University employee. This private distributor became unable to stay current with his bill, and Professor Gregory was concerned about how much money was owed to Purdue from this route. His concern was so great that he had the creamery take over the operation of the route and the account. On April 18, 1932, Gregory sent a letter to Dean Skinner suggesting that the creamery operate the route in the future and asked Dean Skinner if he would approve this action if Controller Stewart also approved its operation. He did not think he would be able to find another private distributor willing to put $2,400 a month into purchasing dairy products from Purdue.[15]

In the late 1920s, the boilers and power plant in Smith Hall were shut down, and the needed power was provided by the university. The power came from the main generating station just north of the Chemistry Building, today known as Wetherill Hall. For years, railroad tracks ran through the grass to the east side of Smith Hall and across State Street, winding its way to the power station. Traffic on State Street was often stopped as a locomotive pushed or pulled railroad coal cars to and from the power plant. In early 1931, Professor Gregory requested the removal of the old boilers from Smith Hall because the extra room was badly needed.[16]

In February 1931, with budgets already tight because of the depression, a crisis loomed within the creamery. The creamery's coil cream pasteurizer, essential for the production of butter, failed because of old age and excessive use. This was an emergency situation because it threatened the continuation of butter production.[17] Gregory sent a letter to R. B. Stewart, university controller that read:

This is a piece of equipment we must have if we want to continue to make butter. $1,700 is needed to purchase a comparable modern vat. Our University budget can stand $500 of this expense; the question is where can we get the remaining $1200? I would appreciate hearing from you with suggestions for this must be taken care of at once.[18]

No answer to Gregory's dilemma has been found, but it is fairly obvious that the money was found somewhere because creamery production continued. Shortly after this problem was corrected, other problems began to pop up. In the latter half of the 1930s, the creamery began replacing much of the equipment purchased ten to fifteen years earlier. In a letter to Controller Stewart, H.W. Gregory requested the purchase of a flash pasteurizer for $900, a buttermilk and starter vat for $900, cream scales for $300, and a modern and larger bottle washer for $1,800.[19] Of these items, the cream scales in use at that time had been condemned by the City Weights and Measures Department. Gregory also wanted to mechanize the bookkeeping, which would lead to a $900 savings per year. All of this was to be paid for out of creamery reserve money that had built up over the many profitable years. Stewart passed this request on to President Elliott with his recommendation for approval. These purchases were approved by Elliott. At a later time, Gregory also requested to replace a cream holding vat because the old one was unusable. The cost of this purchase was $1,400. Similarly, Gregory requested an ice breaking machine, valued at $800. The old ice breaking machine had been purchased in 1912, when ice was manufactured in very large and heavy blocks. These items were proposed and approved to be purchased at a later date.[20]

One of the more unusual budget requests appeared in the 1936–1937 request in which Gregory asked for $1,500 to build a truck garage. His note to Dean Skinner read:

In carrying on the business of the Purdue University Creamery, it is necessary for us to operate a number of trucks, and it is necessary at the present time for us to rent garages in the city for these trucks. It is very desirable that we have our own garage for our trucks.[21]

This reference to storage of the creamery trucks away from campus was the only one that I found while researching the creamery's history. The request was apparently turned down as no evidence of a garage has been found.

Creamery loading dock and back side of Smith Hall.
(Courtesy of Purdue University Archives and Special Collections, Purdue University Libraries)

In June 1938, an additional $9,000 was allocated to purchase heating and pasteurization equipment and to build partitions within certain rooms of Smith Hall. The money for these changes also came from the creamery reserve fund and a special W. C. Smith Fund, which held money from the original donation in 1912 toward the construction of Smith Hall and the creamery.

7
World War II and the Creamery

In the summer of 1941, a tile floor was installed in the creamery's butter room.[1] This was the last of the improvements and replacements in the creamery until after the Allied Victory in World War II. When the war started, most materials and resources throughout the country went into a successful war effort. Even thirty percent of the butter the Purdue Creamery produced was directed to the military or the Federal Food Distribution Agency.[2] As during the First World War, Purdue again installed a program in Dairy Manufacturing for female students to support the war effort during World War II. The Dairy Husbandry Class 114, "Problems of Dairy Technology," was designed with a special section created specifically for home economics students.[3] This class was offered during the second semester of 1942 and was taught by Professor Gregory and Mr. Brown. The class comprised one lecture hour and four laboratory hours each week, which was the equivalent of two and two-thirds semester hours of credit. No enrollment figures have been found, and it is unclear whether this special section was continued for the later war years.[4]

In late May 1943, Professor V. C. Manhart of the Dairy Department, along with female members of the creamery staff, began conducting consumer preference studies of different butters that the creamery could produce. Again, with World War II causing food shortages, Purdue was trying to get the most out of a shortage of available butter. Three local Piggly Wiggly Grocery stores joined forces with Purdue and conducted sampling surveys with consumers. These Piggly Wiggly stores were at Ninth Street and Kossuth, Fourteenth and Howell, and at a Northwestern Avenue location. The ladies of the creamery spent the mornings at the creamery preparing the butter samples and then were at the grocery stores from 2 p.m. to 6 p.m. Customers were asked to taste different butter samples and tell the Purdue testers which samples they liked best.

Professor Manhart even sent a letter to Director W. V. Lambert requesting that he stop by one of these locations and sample the butter himself. Director Lambert was the associate head of the Agricultural Experiment Station.[5] A copy of this letter was also sent to Dean Reed.

By February 1943, the lack of workers caused by World War II that was affecting the country was also affecting Purdue. Both the Purdue Dairy Farm and the Purdue Creamery were short of employees. In a letter from Professor Gregory to Director Reed, Gregory identified these manpower deficiencies. Draft registration during this timeframe was for men ages 18 to 38. Certain wartime jobs were deemed as vital or "essential" by the federal government to keep the country operational and for military supplies to continue unabated. These jobs, if filled by a man, could exempt that individual from military service. The civilian activity of these workers had to be designated as necessary for national defense or as essential farm labor. Because of this issue, Gregory also requested Reed's ideas and input on trying to get the farm and creamery positions labeled "essential." This was a question that men younger than 38 years of age had been asking Gregory.[6] The answer to this question to Read remains a mystery.

The shortage of labor during World War II was hitting not only the Purdue Creamery but also creameries throughout Indiana. Huge numbers of trained men in the dairy industry were volunteering or being drafted into military service. In November 1942, Associate Dean Freeman anticipated this situation in a letter sent to President

Elliott. Freeman stated how he thought enrollment in the 1943 Winter Course for Dairy Manufacturing classes would be very low if a student had to be there for an entire eight-week session.

Upon consultation with dairy executives and plant managers, Freeman suggested offering blocks of one-week classes instead. In this way, a student could register only for the training he was interested in obtaining and be in classes for just that period of time. The student could register for one, two, three, or however many sessions he desired. The student might be on campus one week, be gone for a period of time, and then return to Purdue for another week. The classes that Freeman wanted to offer were Market Milk, January 18–23; The Manufacturing of Dairy By-Products, January 25–30; Ice Cream Making, February 1–5; Butter Making, February 8–13; Cheese Making, February 15–20; Dairy Plant Mechanics, February 22–27; Milk and Milk Grading, March 1–6; and Dairy Technicians' Course, March 8–13.

Freeman asked for immediate approval so announcement of the changed plans and timetable could be given to the dairy manufacturers of the state. Freeman's final sentence of the letter underscores the seriousness of the war effort when he writes, "This proposed plan involves no great cost but does present a modification designed to prove helpful in times of war stress when trained technicians in dairy manufacturing industries are very seriously needed."

This plan was apparently approved by President Elliott even though official notification cannot be found. Handwritten at the top of this letter in green ink is a note that says "O.Ked Pres Elliott, Nov 21/42."[7]

The failure of this modified class schedule is probably not open to interpretation because not a single student enrolled. Nine students enrolled in the dairy production coursework and were given an opportunity to complete the manufacturing classes. The April 16, 1943, minutes of the Board of Trustees reported, "Dairy Manufactures was offered on a one week unit basis. No student on this basis completed all the training."[8] For comparison, it is known that in 1940, Winter Course Dairy Manufacturing graduates totaled twenty-five. Ten months prior to the Pearl Harbor bombing, the 1941 number had dropped to thirteen. In 1942, the number was nine, for 1943 ten, in 1945 just four students, and in 1946, immediately following the war, the number of graduates was at fourteen. It is known that for the 1944 graduates, the schedule had returned to the more traditional eight-week classes.

The announcement for the 1944 winter classes once again extolled the handicaps the dairy manufacturers of Indiana were facing because of manpower shortages and federal government demands, as shown here:

> *Dairy plants are confronted with innumerable procurement, plant and marketing problems related to the war effort. These current problems must be met with inadequate, untrained, and inexperienced labor.*
>
> *The demand for high quality dairy products for a government food reserve has focused attention upon farm and plant sanitation, grading raw materials, scientific processing, and judging of the products.*
>
> *Scientific methods of manufacture must be based upon modern control tests to secure the composition specified by legal definition.*
>
> *Quotas must be established and formulas adjusted to meet the restrictions governing production and sales.*
>
> *Elaborate plant records must be kept as a basis for government reports. The increased demand for dairy products requires more rapid and continuous operation of equipment. Programs of instruction to conserve equipment, material, and supplies are necessary.*
>
> *Increased costs of raw materials, supplies, labor, and other items of overhead must be met by increased efficiency. Plant efficiency is based upon trained and productive labor. Only trained operators are qualified to assume the present responsibilities in the key positions.*
>
> *The essential training required is offered in the Eight-Weeks Dairy Manufacturing Course. This program is planned to furnish the fundamental information and develop the necessary skills to make the individual valuable in a dairy plant. Thirty-five hours of instruction covering seven subjects are given each week .*[9]

The urgency for maximum production and maximum efficiency resulting from highly trained but minimal amounts of labor was felt by all of American industry, not just the dairy industry during this time of national crisis. The announcement for the 1945 Winter Classes was even more direct and blunt in regard to available labor and the quality of the remaining workforce available to the Indiana dairy manufacturers.

> *The urgent demand for a huge quantity of high-quality dairy products cannot be met by the few trained survivors in dairy plants who must rely upon inadequate and untrained help.*
>
> *These key men are confronted with many new problems related to the war effort. Their attention is focused upon farm and plant sanitation, volume production, grading, scientific processing, and the control of composition. Quotas, formulas, records, and sales must be adjusted. Equipment, materials, and supplies must be conserved. Increased costs must be met.*
>
> *The present situation demands increased efficiency based upon more labor that is adequately trained.*
>
> *Each manager should select a man for short-course training this year. The Dairy Manufactures short course is planned to furnish the student fundamental information, multiply his experiences, and broaden his services so that he will be a more valuable operator.*
>
> *There are special opportunities in dairy plants for those who take time to train for the work. Such training will not be lost, because the demand for dairy products will not disappear when the war ends.*[10]

In 1944, there was discussion between Dean Reed, Associate Dean V. C. Freeman, and Professor Gregory about building an enclosed loading dock onto Smith Hall and the Purdue Creamery.[11] They decided that, because of other proposed improvements for the creamery, the construction of the enclosed loading platform should be delayed. In a letter to Dean Reed, Associate Dean Freemen said, “I think it would be a mistake on our part to spend several hundred dollars on this kind of an improvement with our other projected plans underway.” What these future plans were have not been discovered and remain a mystery. H. T. Hamilton, superintendent of the Purdue University Physical Plant, was also seeking approval to cancel a roofing contract on the now defunct enclosure.[12]

8
Explosions, Accidents, Road Dust, and Sour Milk

Elliot H. Parfitt's name is seen in the course descriptions as the teacher for several of the previously listed classes. His name and works kept appearing in articles researched for this writing. Parfitt, although not directly involved with the creamery, was a noted contributor to the fine reputation Purdue earned in dairy investigations and dairy product manufacturing. His tenure at the university started in 1921. He earned national attention for his research of bacteria involved in dairy products and milk proteins. He was also noted for his experimentation in chemical sterilization in the dairy industry. His research on the cleaning and sterilization of on-the-farm milking machines eventually led the Indiana State Board of Health to adopt those practices into procedures approved as standards for the State Board of Health.[1]

In the 1936–1937 Dairy Department budget request to Dean Skinner, Professor Gregory asked Skinner for a $200 increase in salary for Parfitt. In his letter, Skinner said:

> *Dr. Parfitt is a splendid teacher and due to an increase in undergraduate and graduate students in the Dairy Department*

> *he is devoting half time or more to teaching work. He is also on several committees relating to the University, all of which requires time. He has been connected with Purdue since 1921, and has received his Master's Degree and Doctors Degree since coming to Purdue. It is very desirable that we raise his salary if we expect to keep him at Purdue, with his ability and training.*[2]

Parfitt resigned from Purdue effective January 1, 1940.[3]

Parfitt was involved in one of the few accidents ever recorded taking place in Smith Hall. The *Purdue Exponent* reported on its front page that Parfitt, who was in charge of the bacteriological laboratory, and students C. C. Walts (class of 1924) were unhurt as an explosion occurred in the laboratory room. The windows of the laboratory and a few others throughout Smith Hall were blown out because of the blast. It was lucky that this accident occurred during Christmas break (December 30, 1922) because classes were not in session and few people were around to be hit with flying glass. The cause of the explosion was a defective boiler used to create steam. An autoclave sterilizer connected to the boiler was destroyed as well.[4] Apparently this accident did not deter Mr. Walts from his education in Dairy Manufacturing as he went on to do graduate work at The Connecticut Agricultural of College.[5]

Student laboratory for milk and cream testing.
(Courtesy of Purdue Animal Sciences Department)

While major accidents were few, small injuries popped up fairly regularly in the creamery. Few reports were ever discovered, but five accidents were reported in just the three weeks between September 8, 1941, and October 2, 1941. Although none of the injuries were serious, two of the reports were for steam and hot water burns. The third accident required stitches, the fourth one was caused by someone slipping on a wet floor, and the fifth one was the result of a hand being caught between a falling milk can and the floor of a truck.[6]

Purdue University has always been divided geographically by State Street because it runs through the campus from east to west. The buildings in the section south of State Street have primarily and traditionally been devoted to agriculture. This is no longer as true in the twenty-first century as it once was. Although we do find new agricultural buildings, such as the new state-of-the-art Creighton Hall of Animal Sciences, in more recent years, there has been a surge in construction of some of the most modern and technologically advanced buildings for the advancement of the nonagricultural sciences at Purdue. We now find the Birck Nanotechnology Building, the Bindley Bioscience Center, the Jischke Hall of Biomedical Engineering, the Drug Discovery Building, and several others.

But in June 1923, State Street itself was far different than it presently is. In the earlier time, the thought of a four-lane thoroughfare had not yet been conceived. Traffic moved much more slowly with far fewer cars and trucks. In fact, in 1923, State Street was nothing more than a dirt road. The dust and dirt kicked up by traffic on this dirt road interfered with the cleanliness and sanitation of laboratory work and creamery production at Smith Hall. This was particularly true because, at that time, Smith Hall was without a cooling system beyond merely opening windows to create air movement. At the June 1923 Board of Trustees meeting, University President Edward C. Elliott recommended oiling State Street in front of Smith Hall. This action would help prevent the dust from the dirt street from becoming a more serious problem for the creamery. In fact, President Elliott referred to the dust as a "menace."[7]

In addition to President Elliott's concerns about dust from State Street, the board was also concerned about the daily operations of the creamery. In 1924, the attic of Smith Hall was filling up with correspondence, cancelled checks, receipt books, and other records stored on paper. The board deemed this collection a fire threat and ordered the destruction of records from 1908 to 1920.[8]

In late 1925, Gregory noticed the wear and tear on classroom space within Smith Hall. He contacted C. D. Bushnell, head of Buildings and Grounds, by phone and asked for repairs to be made prior to the start of the Winter Course in Dairying. Room 112 was a particular concern as it was becoming almost unusable. Mr. Bushnell was slower to act than Gregory wanted, so he asked Dean Skinner for help in order to facilitate speeding up the necessary repairs.[9]

Concern for the upkeep and appearance of the creamery focused not only on the interior but also on the exterior. In April 1928, Professor Gregory wanted to upgrade the shrubbery and replace plants surrounding Smith Hall that had been damaged or destroyed by the winter weather. He also wanted the road that came in behind the creamery to be widened and covered with fresh gravel.[10]

Dean Skinner also observed Smith Hall trash being left around the building, the Agricultural Campus, and the entire university, particularly leftover ice cream containers strewn on the floors and grounds. He encouraged Gregory to put up some signs in the creamery salesroom and Smith Hall requesting students to put trash in the proper receptacles. This was an ongoing problem over many years, as Skinner noted to Gregory many times in letters.[11]

Skinner also sent a letter to Gregory about creamery employees and Dairy Department staff parking their cars along the driveway into and out of the loading dock area of the creamery. Parking in this area made it difficult for trucks to get into and out of the creamery. Skinner suggested putting up "No Parking" signs and recording the license plate numbers of offenders.[12]

The creamery occupied the time and thinking of Dean Skinner on both a professional and personal level. Dean Skinner was a patron of the creamery and had dairy products delivered to his house by a regular route driver. As a customer concerned about quality and value, he was not afraid to complain when his milk was not up to standards.

For example, the first week of July 1927 was very hot, and apparently milk delivered for two consecutive days had soured, and "was not in first class condition" on two other days. He even personally called Professor Gregory on the phone one evening to complain. In a letter, Skinner suggested that Gregory check with the delivery route driver to see he if had other complaints.[13]

As noted in the previous pages, Gregory was constantly dealing with problems, concerns, and complaints. Traffic accidents were also a topic Gregory had to deal with in his career as Dairy Department head. Harold Kantz was known to be a route driver for the creamery at least during 1938 and 1939 and possibly other years. Kantz was involved in a slight collision with a train in Lafayette as he was running his home delivery route.[14] In the January 18, 1939, meeting of the Board of Trustees, a letter from Gregory was read concerning a traffic accident involving a Purdue Creamery delivery route driver. Apparently, on December 10, 1938, at 10 a.m., Kantz collided with a Wabash Railway freight train at 6th and Romig Streets in Lafayette. It was reported that Mr. Kantz did not see the train until it was too late, and he wound up hooking the right front fender on a freight car. This accident damaged the fender, the head light, and a tie rod, causing $40 in damage. Mr. Kantz had the truck towed to a local repair shop, obtained another delivery vehicle, and finished his route.[15]

9
Employees of the Creamery

Finding the names of creamery employees has proven to be exceptionally difficult. The earliest known employees were all listed as butter makers. These were the foremen or crew chiefs of the creamery, the most important men in the process of making butter and of manufacturing other dairy products. Without an outstanding butter maker, creameries often struggled to obtain financial success. With butter sales spanning much of the Midwest and East, it is apparent that Purdue hired very qualified personnel as butter makers.

Butter maker H. N. Slater was previously mentioned as being hired by Professor Hubert Van Norman as early as 1903. The 1905 *Debris* yearbook gives a brief description of Mr. Slater:

> *"Pa Slater"—Yes that is what we call him. He is the genial butter-maker and ex-center on the Minnesota foot ball [sic] team. He appreciates a good joke so well that it would be a shame to tell one on him. Though not an orthodox professor, all will agree that he is the best all-around man on the place and even Van has already found out that it does not pay to buck such excellent foot-ball [sic] material, for though not in training, he is reinforced by years of work with cream and butterfat.*[1]

On June 30, 1906, Slater left Purdue and was employed by the U.S. Department of Agriculture, receiving a higher salary there.[2] Slater died sometime prior to the spring of 1912. A statement about his passing was included in the 1912 *Debris*.[3]

H. J. Pellam was listed as Purdue's butter maker as early as August 1911.[4] He was a student at the university in 1906 as evidenced by the following excerpt from the *Purdue Agriculturist*: "It is gratifying to notice the dairy training offered at Purdue is such practical value that her students can make a showing in competition with other dairymen. H. J. Pellam won the first prize cup in the gathered cream and gathered butter class in the educational scoring contests for the past year."[5] This was a series of six contests, one every two months for butter makers to test their skills in producing quality butter.

Harry Damen Smith was an assistant butter maker for the creamery at the outbreak of American involvement in World War I. He was required to register for military service as were many other School of Agriculture employees. This was a major concern for Dean Skinner and other members of the Purdue administration.[6]

The creamery budget request for 1929–1930 listed Harry Nicely, Archie Murphy, Ida Earhart, George Kantz, Vola Widner, and Owen Richardson as "Service Staff." The "Clerical Staff" was made up of Dorothy Campbell, Irene Deepwell, and unlisted part-time help.[7] By the budget year 1930–1931, Ida Earhart had left the employment of the creamery and Francis J. Richardson had been added. Richardson was listed as "floor foreman." Dorothy Kress replaced Dorothy Campbell on the clerical staff.[8]

Of note here was a letter sent from Gregory to Skinner on December 17, 1931. It stated, "It was the Purdue Creamery policy to employ only young men who have completed dairy manufacturers short course or who have had training in dairy work."[9]

Harry Nicely split his time between being the building engineer for Smith Hall and for the creamery. In the 1937–1938 budget request, Professor Gregory asked that $90 be returned to Mr. Nicely's annual salary to bring it up to $840. Apparently, this portion of his salary had been removed because of the money crunch brought about by the depression. In this request, Gregory called Nicely "a very valuable man."[10] Mr. Nicely stayed on the job until an unknown date in 1944–1945.[11] The reason he left this employment is also unknown.

L. C. Murphy replaced the departing Nicely in 1946.[12] The next person found to be employed as "Building Engineer" was Everett Johnson, who, like Nicely, split time between duties in Smith Hall and the Purdue Creamery.[13] It is known that Everett Johnson replaced Murphy effective December 1, 1951.[14] Johnson was still performing these duties as late as 1956.[15]

E. A. Siebenthal was a 1931 graduate of the Winter Course in Dairy Manufacturing and was a deliveryman for the creamery in October 1937. Purdue required a surety bond for his employment.[16] The surety bond guaranteed Purdue the payments it was due from Mr. Siebenthal. Mr. Siebenthal was possibly purchasing dairy products wholesale from the creamery and reselling them on a retail delivery route.

Marshall Johnson was employed by the Purdue Creamery for only a short time, but received many injuries doing his tenure. In a letter to President Elliot on October 11, 1943, Mr. R. B. Stewart discussed Mr. Johnson's injuries, disabilities, treatment, and cost to the creamery.

> *This will report upon the disability of Marshall Johnson, an employee in the University Creamery who is being given medical treatment in accordance with the Workman's Compensation Law and as directed by the Board of Trustees for such cases. Mr. Johnson was hired by the Purdue University Creamery on April 8, 1941, where he has worked at various activities since that time.*
>
> *It is reported that he figured in an accident on November 9, 1941, when he slipped while lifting a can of milk from a truck to the receiving platform, at that time tearing loose eight vertebrae. Dr. Miller treated this injury. On January 17, 1942, Mr. Johnson again injured his back while lifting an ice cream packer, and Dr. Miller again treated the employee.*
>
> *On October 30, 1942, he was rather severely injured when two cases of milk fell on him while he was working in the milk storage room. As I understand it, Mr. Johnson was not responsible for this accident. At this time, the employee was referred to Dr. E. T. Stahl, under whose care he has been receiving treatment from time to time since that date. He has spent several weeks in the hospital at Creamery expense and at the present time it seems that his condition is not improving. We have been reluctant to advise surgery in the case of this employee and feel definitely that aside*

> *from carrying it forward under the Workman's Compensation Law until the necessary total payments and medical care have been provided, the University will be wise to allow Mr. Johnson and the Doctor to make their own determination of treatment.*
>
> *Up to the end of September, 1943, the case has cost the Creamery a total of $1,574.36 for the last injury only. According to the law, the University must continue medical treatment and pay 55% of Mr. Johnson's average weekly wages of $18.00 until returned to work, or 300 weeks.*[17]

On October 9, 1944, Mr. Stewart informed President Elliot on the final disposition of this case. Mr. Johnson received 25% total permanent disability payment of $2,037.08, and Purdue paid an additional $1,259.48 in medical costs.[1]

Others known to have been employed were Bernice McMurtrie, listed as a member of the clerical staff in 1941–1942.[19] She was again listed as a staff member for 1944–1945.[20] Robert "Red" Oilar was a home deliveryman employed by Purdue for ten years.[21] R. A. Brown was reportedly a home delivery route driver in January 1948. The duration of his employment is unknown.[22] Other employees were Ralph Johnson,[23] Gary Anthrop,[24] and Les Martin.[25] Also found were the names of John Dienhart for employment in the creamery and of Maxine Beeker as temporary clerical help.[26] John Byers was employed for creamery help in October 1943.[27]

James Roscoe "Rocky" Hurst worked until the creamery closed. Hurst delivered milk and dairy products throughout Tippecanoe County. After the Purdue Creamery closed, he was employed by Dean Foods doing very similar work.[28]

Another person believed to have worked to the end or very near the end was Isaac "Ike" High. High was a route delivery driver and sometimes worked in quality control. He finished his Purdue employment at the dairy farm upon his retirement in 1977.[39]

10
The Years of Rhodes, Babel, and Parmelee

Verne D. "Dusty" Rhodes was hired effective April 1, 1944, as assistant creamery manager. His hiring is significant because Rhodes was later promoted and became the final Purdue Creamery manager; he was in charge when the creamery shutdown in 1969, through no fault of his own. Rhodes was a 1923 graduate of Penn State University and had a bachelor of science degree.[1] Rhodes transitioned to the Purdue Creamery from the Agriculture Experiment Station Creamery License Division as an assistant creamery license inspector.[2] He started in the Creamery License position in 1923, making $1,500 annually.[3] When his employment changed from the Creamery License Division to the Purdue Creamery, his salary rose to $2,550.[4]

Dusty Rhodes. *(Courtesy of Purdue University Archives and Special Collections, Purdue University Libraries)*

Mauri Williamson, a former student of Rhodes, has fond memories of Dusty Rhodes "smacking his lips" when teaching about ice cream or milk tasting and sampling. He recalls Rhodes as being methodical and deliberate. "He was a good teacher for the students," said Williamson while being interviewed at the Pioneer Village during the 2014 Indiana State Fair.[5]

Even though most of the post-war era discussion of the Purdue Dairy Department and Creamery centered on the planning and construction of a new dairy farm located northwest of the campus on Cherry Lane, there was still discussion at the May 10, 1945, Board of Trustees meeting concerning an addition to Smith Hall, which was estimated to cost $300,000. At this time, the board did not fund the addition.[6] In April 1947, they sought the authority of the State Budget Committee to make the additions.[7] Revised plans were then drawn up because the Life Science Building was to be erected immediately to the west of Smith Hall.[8] The Life Science Building plans were already put together. The bids for the Smith Hall additions were then submitted and approved on October 20, 1948, with Kuhne-Simmons Company having the low bid of $654,000 plus $120,500 to cover the architects' fees, equipment costs, and other expenses.[9] The one-story east and west wings built onto Smith Hall provided brand new office space for dairy production personnel and a laboratory for feed and dairy research analysis.[10] Basements under each wing were also constructed for additional storage space, additional restrooms, and showers.

It was after these new additions of the east and west wings that Dr. Will Dillon witnessed one of the more humorous incidents he remembers happening at Smith Hall. Dillon, a dairy extension specialist, had his office located on the second floor of the original section of Smith Hall. The east and west additions were only one story tall, and there was a door on the second floor of the old part of Smith Hall that led to the roof of the new east wing. One day when Dillon was in his office, he heard a commotion on the roof, and when he looked, he saw a handful of students stuck on the roof. The curious students had begun exploring the roof when the door to the roof locked behind them. Some of these students thought about jumping to the ground but reconsidered when they saw how far the fall would be. After watching them for a couple of minutes, simply for the entertainment value as they stewed in their predicament, Dillon finally acted. The explorers were rescued when Dr. Dillon reluctantly opened the door for the embarrassed young men.[11]

In 1948, the university charged the creamery $341 for public liability insurance.[12] The university apparently felt that it needed to increase its liability coverage throughout campus. This decision was spurred by the collapse of bleachers at a Purdue basketball game. This collapse in February 1947 inside Lambert Fieldhouse led to the death of two people and injuries of scores of other fans.[13]

Not until the early years of the 1950s was there much new activity in regard to physical upgrades and remodeling of the creamery. In April 1950, fluorescent lights were installed in Smith Hall.[14] In January 1951, $7,700 was allocated from creamery funds to pay for additional repairs. Additionally, $76,000 was spent in December 1951 for rebuilding the receiving and storage section of the creamery.[15,16] The contract for this job was awarded to A. E. Kemmer at an actual cost of $68,670.[17]

On January 1, 1954, Ralph Roberts turned over the management of the creamery to Dusty Rhodes.[18] Roberts had been hired as the manager of the creamery in 1924, although the exact starting date was not found. Although Tuttle resigned on November 15, 1924, the Purdue Board of Trustees meeting minutes do not reveal details about the hiring of his replacement, Ralph Roberts. After handing over the management of the creamery to Rhodes, Roberts stayed with the university until his retirement June 30, 1965.[19]

In the 1953–1954 budget requests to Dean Reed, Professor Gregory requested a pay increase of $480 per year for Rhodes. Gregory knew that Purdue was receiving a bargain with Rhodes' salary and that it would cost Purdue considerably more to replace him.

> *Professor V. D. Rhodes is now Assistant Professor. I am recommending that his title be Assistant Professor in the University and Superintendent of the Purdue Creamery. He handles all labor and the operation of our Creamery. Professor Rhodes is an excellent teacher. He is very practical and appeals greatly to underclass students and short course students. Professor Rhodes is also a keen business man [sic] and his advice is often sought by graduates entering commercial work. To replace Mr. Rhodes in instructional work and in our creamery would*

> *take considerably more money than he is being paid. I am recommending that his salary be increased $480 on University funds.*[20]

Gregory asked for this pay increase through university funds because the dairy department funds for payment of personnel could come from as many as three different sources: university funds, creamery funds (from the profits of the creamery), and agricultural extension funds.

Rhodes' salary ended up being increased from $2,900 to $3,500 when he was promoted from assistant creamery manager to manager at the start of 1954.[21] Rhodes was firmly in charge of the creamery and its staff. Many staff members caringly referred to him as "The King in His Castle."[22] He was also an individual who wanted to stay attuned to the economic situation in the United States and in the world, and was seen reading *The Wall Street Journal* on a daily basis.[23]

Perhaps more important than the physical improvements to Smith Hall and the personnel changes during the 1950s were the scientific discoveries and improvements in dairy products by the scientists and employees of the creamery. Dr. Fred J. Babel, hired in 1947, directly produced some of the most significant developments in Purdue dairy foods processing.[24] His brilliance led to patents, fame, and worldwide respect in the dairy manufacturing industry for both himself and Purdue. For example, Dr. Babel developed a cream dressing that was added to cottage cheese to retard bacterial growth, lengthening the shelf life of the product.[25,26] During this time, the Purdue creamery was manufacturing between 4,000 and 4,500 pounds of cottage cheese per month. The cottage cheese was the only type of cheese Purdue was selling commercially during the late 1940s and early 1950s.[27]

Dr. Fred Babel. *(Courtesy of Purdue University Archives and Special Collections, Purdue University Libraries)*

Other achievements of Babel's experimentation and research were the identification of important activities on the dairy farm and in the milk house to ensure that the raw milk was produced and handled so as to achieve and maintain a high-quality product.

These efforts primarily focused on sterilizing and sanitizing milking machines and related equipment. After identifying these activities, he suggested replicating many of these same procedures in the handling of the raw milk at the dairy product processing plant. In 1957, Dr. Babel coauthored the textbook "*Dairy Bacteriology*" with David Hammer. This text was still being used extensively at the time of his passing.[28]

The American Dairy Association was duly impressed with the scientific advancements at Purdue, so they paid the university $6,000 to improve the flavor and shelf life of sour cream dressings in 1964.[29] They awarded another $6,000 to continue this research in 1965.[30] The ADA again paid $6,000 to Purdue in March 1967 for the development of sterile canned mixes for soft serve ice cream products and milkshakes.[31] In the late 1950s, Dr. Babel was also credited with the development of the famous Purdue Swiss-style cheese.[32] This cheese is still sold by the Purdue Agriculture Alumni Association, and the proceeds fund student scholarships, although the cheese is now manufactured by a non-university vendor. Dr. Babel won the recognition of the American Dairy Science Association when he received the Pfizer Award in 1961 and the Kraft Teaching Award in 1973. He retired from the university in 1978 as professor emeritus of Animal Science.[33]

Even though Dr. Babel stayed at Purdue his entire career, Purdue had been uncertain about whether he would remain at the university. In a letter from Professor Gregory to Director Reed, Gregory expressed both his concern for possibly losing Babel and regret at the low pay Babel was receiving from the university. It is unclear what his level of pay was in 1949, but Dr. Babel had been offered more than $10,000 per year to enter private industry. He had offers from both The Ohio State University and the University of Illinois, with each of these sister Big Ten Universities offering more than $6,000. In the letter to Director Reed, Gregory recognized Babel's talents in the classroom and the laboratory, calling him "one of the outstanding young dairy bacteriologists in the country." Gregory ended his letter to Reed with, "It would mean a great loss to us if he would happen to leave the Institution which may happen with the increase which we have given him."[34] A tribute to his work and life was printed in the 1978 summer edition of the *Purdue Alumnus*:

Ag Alumni Cheese Inventor Retires

What's white, either small or large, and the all-too-familiar friend of anyone who has ever been on a diet?

What's creamy white, holey, comes in big wheels, and is known worldwide as a taste delight?

And what do these two things have to do with Purdue?

The first answer is cottage cheese and the second, the famed Ag Alumni Swiss-type cheese. And they represent milestones in the distinguished career of Dr. Fred J. Babel, (MS36) professor of animal sciences, who this month wraps up 31 years on the Purdue faculty.

Babel is the "father" of Ag Alumni Swiss Cheese, which for many years has had an international reputation as a gustatory delight. Known for its soft, creamy texture and its distinctive nutlike flavor, Purdue's Swiss cheese had unlikely beginnings as Babel's Ph.D. thesis at Iowa State University.

"My thesis problem was to develop a small swiss{sic} cheese made from pasteurized milk. When I came to Purdue in 1947 we did more research on this type of cheese. It happened that cheese manufactured in the Purdue Creamery was well suited to student's schedules — with vacations and such — a surplus of milk was available, so we began to produce it experimentally in the Creamery.

"Well, then we had to get rid of all of this experimental cheese because Smith Hall was to be remodeled, so we prevailed upon some local outlets to accept two or three wheels to sell. Acceptance was so good that they came back for more immediately. Purdue Cheese was manufactured in the Purdue Creamery until its closing in 1960 [1969]. After the Purdue Creamery closed, the demand for the cheese was still so great that the Purdue Ag Alumni Association decided to try to sell it. The formula was sold to Kentland Dairy in Mulberry who manufactured it until just recently when the firm was sold to Freshmark Foods, Inc.

"The popular cheese is an adaptation of a Danish method for manufacturing Samsoe cheese. When the creamery made it, we manufactured it in small wheels of 6–7 lbs. and in 20 lb. blocks," he said.

"The cheese is made from pasteurized milk, which makes it

different from regular Swiss cheeses, it has a higher fat content and is manufactured at a lower temperature and with less moisture content than standard Swiss," [said] Babel.

Besides being a taste delight, Ag Alumni cheese represents a very real monetary boon for both the Purdue Alumni Foundation and the Ag Alumni Trust Fund, who receive a royalty from every piece sold.

A second stellar achievement in Babel's career was his discovery of a method for improving the flavor and keeping quality of cottage cheese, one which more than tripled the shelf life of the product.

The research began in 1960. Babel was seeking to stabilize and improve the flavor of cottage cheese.

"Cottage cheese's flavor is dependent on the manufacture and type of culture used, and all manufacturers had a problem with wide variation in the flavor quality. We knew the flavor components were derived from citrate — so in attempting to standardize the flavor, we added increased citrate to the milk. So we thought if we increased this, we could increase the flavor components. But, when the whey was drained and the cheese was washed, we lost most of the compounds," he related.

"So the next thing we attempted was to alter the creaming mixture. Creamed cottage cheese that you buy consists of 2/3 curd and 1/3 creaming mixture, which contains 12 per cent [sic] milk fat. So we decided we ought, instead, to form the flavor components in the creaming mixture.

"That didn't work out too well, because we had a sterilization problem. So, next we decided we would culture the microorganisms in skim milk, then if we developed a high flavor in skim milk, we could add that to the cream to get the desired flavor," he continued.

"Well, this worked. And besides that, we found that cottage cheese handled that way didn't spoil so fast as that made by the regular method. It was found to more than triple the shelf life."

"Soon, major dairy manufacturers were expressing interest in the process, and we obtained a patent on it. Today this method of creaming is used by every large manufacturer in the U.S. and Canada, "he said.

"Under this method, a plant could pay Purdue royalties on the

process, and still make money, because there is virtually no loss, no returns, no package losses, etc." Babel said.

So what does a researcher do for an encore? More of the same. Babel in recent years has been working with bacterial cultures in cheese manufacture, fermented milks, etc., the microbiology of dairy products in general, the growth characteristics of these microorganisms and their enzymes.

After retirement, Babel plans to "first, clean out my office, which after 31 years is a major chore," he laughed, "then publish some research which I haven't had time to get written up yet, and do some consulting work."

What will he miss the most? "The students," he says emphatically. "I've had a very heavy teaching schedule, both undergraduate and graduate.

"There has been a tremendous upsurge in the food sciences, I believe because students want to enroll in a curriculum where they can get a position. Food science has been a good curriculum and has shown steady enrollment increase, doubling in the last 5–6 years."

Babel, a native of Michigan, came to Purdue in 1947, after earning the Ph.D. from Iowa State University in 1939. He received his MS [M.S.] from Purdue in 1936 and his B.S. from Michigan State University in 1935.[35]

Dr. Merle Cunningham, former Animal Science professor, remembers his colleague as "always being very professional, always clean and neatly dressed. His office was always clean and arranged with things put away. His laboratory was clean enough to eat off the floor." Cunningham went on to say that Babel had the neatest and most readable handwriting of any male he ever knew.[36]

Dr. Carlton E. Parmelee, a fellow faculty member and close colleague of Dr. Babel, also worked with the Purdue Swiss Cheese. As previously mentioned, Dr. Babel brought the process with him from his Ph.D. work at Iowa State. Both he and Dr. Parmelee continued to experiment and improve the cheese. According to Bob Hessong, a former Dairy Manufacturing student, Dr. Parmelee was known by some of his dairy production students as "The Cheese Professor" during the middle to late 1950s.[37] The following article from the

November 10, 1956, issue of *The Purdue Agriculturist* discusses the work of Parmelee and the Purdue Swiss Cheese manufacturing method.

Dr. Parmelee inspecting Purdue Swiss cheese.
(Courtesy of Purdue University Archives and Special Collections, Purdue University Libraries)

> *Dr. C .E. Parmelee of the Purdue University dairy department has been experimenting with a new process of making Swiss cheese. The process used to make the Purdue Swiss is a modification of the Gouda cheese process with the addition of a bacteria [sic] that produces the eyes or holes in the cheese in a shorter aging period. The true Swiss cheese has a different process of heating and cooking, a different texture, and requires a longer aging period. It is usually made in wheels of 250 pounds and has a thick rind.*

> *The process which Dr. Parmelee has developed for Purdue Swiss consists of a curing process of 6 to 8 weeks to open the eyes and then holding for 4 weeks to develop the flavor. The cheese is rindless due to the special process of curing which produces no waste. The Purdue Swiss type is made in family-size wheels of 5 pounds. The only disappointing factor is that the Purdue Swiss type is not produced in any quantity for commercial sales.*[38]

Purdue Swiss Cheese label.
(Courtesy of John Cleland)

Purdue released a recipe (date unknown) to the public for this cheese in an Agricultural Experiment Station bulletin. Although this recipe is not the exact formula that Babel and Parmelee used, it is slightly simpler, less detailed, and easier to follow. The major difference between the original recipe and the mass-produced recipe in the bulletin is that the latter recipe assumes the use of pasteurized milk. The original recipe instructs the cheesemaker to pasteurize the milk at 161°F for 16 seconds. It also strongly suggests using only high-quality milk with no off flavors and no developed acidity. This pasteurized milk should then be placed in a sanitized cheese vat.[39] Here is the easier recipe:

PURDUE SWISS-TYPE CHEESE
How to Make Swiss-Type Cheese
A cheese with a mild, sweet flavor

Ingredients — Grade A Pasteurized Milk, Lactic Acid Culture, Propionic Culture, Rennet, Salt

Equipment — Cheese Vats, knives, press, strainers, cloths, and hoops
Refrigerated Rooms — 1 60° room, 1 40° room
Brine tank

Cheese Making Instructions

Heat pasteurized milk to 86°F and add 1% lactic acid culture and a pure culture of propionic bacteria. Propionic culture produces the normal sweet flavor of Swiss cheese and carbon dioxide which forms the holes or "eyes."

After the lactic acid culture has produced a slight amount of acid, coagulate the milk with rennet. Next cut the curd into 1/4 inch cubes and stir constantly for about one hour. During the hour, you must remove 1/3 of the whey and replace it with enough hot water to raise the temperature to 102°F. Continue cooking for about a half hour or until the combination of heat and acid reduce the moisture in the curd to a legal limit.

When the cooking is completed, mat the curd into one large block and then cut into smaller blocks that will fit the size hoop you are using. Put the hoops of cheese under slight pressure for 3 hours until the cheese mats into a solid piece which fills the hoop. Now the cheese is ready to be placed in a 26% brine solution for 2 days. After drying for one day, swiss [sic]-type cheese is ready to be packaged for curing.

Curing Instructions

Refrigerate the cheese at 60° for 6 to 8 weeks so the propionic bacteria will grow and produce the carbon dioxide to form the eyes, and the propionates to give the cheese the sweet swiss [sic] flavor. When the eyes are as big as dimes, move the cheese to the 40° storage and hold it for at least 6 weeks so the flavor will develop.

CAUTION! Flavor will continue to develop slowly at 40°, so those of you who do not like a strong cheese had better eat all you make within 6 months.

Ready Made Cheese

Purdue dairy researchers have developed this experimental Swiss-type cheese. To date, it has been marketed only through the Dairy Department Sales Room, but the consumer acceptance studies will soon place a limited supply in several Indiana grocery stores. It is hoped that commercial supplies will be available in the very near future.

This Purdue cheese has the characteristic Swiss flavor but is mild, sweet and has a slightly nutty flavor. It is soft and waxy yet firm enough to slice when cold. Since the cheese is rindless, it is entirely edible.

Purdue Swiss-Type Cheese is nutritious too, [and] a one ounce serving supplies one-third of the daily requirement of calcium, one-fourth the daily requirements of Vitamins A and B_2 and one-tenth the daily protein requirement of a grown adult.[40]

Swiss cheese production by Purdue personnel during the early and experimental years of the cheese was quite erratic. Production showed wild fluctuations without any trends from 1954 to 1960. In 1960, nearly eight times the cheese was produced as was manufactured just two years earlier.

SWISS CHEESE SOLD, POUNDS/YEAR						
1954	1955	1956	1957	1958	1959	1960
8,686	11,015	4,185	4,739	2,408	5,477	18,861

The Purdue Dairy Club started shortly after Smith Hall opened to help promote the Indiana Dairy industry, support the dairy faculty, engage student educational activities, and encourage friendships between people of similar backgrounds and interests. The first Dairy Products Judging Team representing Purdue University was formed in 1918, and the first dairy cattle judging team was formed in 1920.[41] In 1963, the Purdue Dairy Club became involved in the production of Purdue cheese, using it as a money-making endeavor. The club packaged and sold Swiss, Gouda, and Cheddar cheeses around Thanksgiving and Christmas time. The club purchased large wheels of these three Purdue cheeses from the creamery and then cut them, wrapped them in plastic wrap, heat-sealed them, and placed them in boxes and shipped them. The selling price of these finished gift boxes through the creamery retail room was $1.85. Each box contained one-half pound of each of the three cheeses. During the 1966 holiday season, 250 boxes were sold, and 750 boxes were sold in 1967. According to reports, the low volume of sales in 1966 was due to a lack of cheese production and its availability to the club.

Purdue Gouda Cheese label.
(Courtesy of John Cleland)

In 1967, the organization began placing the three types of cheeses, still cut in half-pound blocks, into colorful tin containers with green cellophane bunny grass in the bottom. The price was raised to $2.25. The profits from these sales were used to support the Purdue Dairy judging team and to fund educational trips for the club members.[42]

The Purdue Ag Alumni Swiss Cheese continued to sell very well even after the closing of the creamery. Dusty Rhodes became associated with the Kentland Dairy Products Company of Mulberry,

Indiana, and production of the cheese continued there.[43] The first year following the shutdown of the creamery, more than 11,000 pounds were produced, netting $1,780 for Purdue scholarships.[44] The next year, more than 24,000 pounds were produced, and $2,300 was raised for scholarships.[45] The third year of production raised more than $3,000 dollars for Purdue student educational needs.[46] However, the number of pounds of cheese produced was not found.

More recently, 12,193 pounds were produced in 2011, and $5,792 was contributed toward Purdue Agricultural student scholarships. In 2012, 11,768 pounds were produced, and $5,884 was raised; and in 2013, 13,442 pounds were manufactured, and $6,721 was raised. The final year for which numbers could be found was 2014 when 13,933 pounds were produced and $6,967 was raised for scholarships.[47]

11
A New Creamery?

In October 1958, Professor H. W. Gregory sent a letter to Dean Earl Butz, Assistant Dean Verne Freeman, and Agriculture Experiment Station head, N. J. Volk. This letter detailed the advantages of maintaining a creamery at Purdue. Many bits and pieces of Purdue University Creamery history are revealed in this letter:

> *Attached is information relating to the operation of the Purdue University Creamery and a statement as to the value which we attach to this portion of the Dairy Department. We believe we have one of the best set-ups for teaching, research and extension work in the dairy industry of any agricultural college, and many people from other colleges have favorably commented on our organization here at Purdue. All branches of the Department through staff meetings and department committees, know some of the problems of our teaching, research, extension, Creamery License, official testing and quality work, [for] which different members of the Dairy staff are responsible. Our men working directly with the Industry in extension, inspection, official testing and quality work are conferring weekly with our research and teaching staff on the problems they have in the field.*

It is very desirable that we continue to increase and improve the work done by our staff by providing a modern addition to our building and modern equipment. At the time the two wings were built in 1950 onto Smith Hall we requested additional space for research, teaching, and a new creamery. It was agreed that this was very necessary by the authorities. Due to the lack of funds after estimates were obtained it was necessary to reduce our request and we were requested to do this. I requested that we be allowed to eliminate rather than reduce the request for teaching and research and I requested that the creamery be eliminated and provided at some later date. We were assured that this would be provided later, in the next few years. It has now been 8 years since this addition was built.

I hope it is still possible after you have looked over the attached statements regarding the operation of the Creamery to come over to the Department and go over the Purdue Creamery and its operation, and confer with the members of the staff who use the creamery in their teaching and research work and who realize the advantages of having a modern Dairy plant.

The American Dairy Science Association appointed a committee to study methods used at several of the agricultural colleges in giving laboratory instruction in the manufacture of dairy products. While there were many points considered in which there seemed to be no general agreement, there were, however, several points which the committee thought could be clearly established and the committee recommended that the policy of the American Dairy Association on the operation of creameries by universities and colleges for the purpose of laboratory instruction be established as follows:

1. *Laboratory instruction in dairy manufactures must be conducted under conditions closely resembling those existing in a commercial plant, if it is to adequately round out the course of instruction.*
2. *The completeness with which factory conditions are imitated should be determined by needs of the industry in each state as seen by the institution concerned.*

3. *Where an institution cannot meet the cost otherwise, it is justified in operating a self-supporting plant, buying and selling on the open market.*
4. *The extent of operation should not go beyond the point where sufficient volume is handled to make the operation of the unit safe from the business standpoint.*

If we did not operate the Purdue University Creamery in order to give laboratory instruction under conditions closely resembling those existing in a commercial plant it would be necessary for us to do one of two things:

1. *Make arrangements for use of local dairy plants as laboratories;*
   a. *Such an arrangement takes the student off the campus which requires expense, time, and places our instructional work under conditions not in control of the university.*
   b. *It would be very difficult to find one plant that handled all dairy products; therefore, arrangement would have to be made in this locality with at least three different plants in order to give instruction in all dairy products.*
   c. *Laboratory instruction under such conditions would by necessity be reduced to a minimum and under such arrangements the number of students who come here for training would be greatly reduced, or reduced to such small numbers that the expense of offering such a course would probably be prohibitive.*
   d. *While under such arrangements the Dairy Department would be relieved of a large amount of worry and problems connected with the operation of a university creamery; however, I believe there would be far less problems to the department personnel under our present conditions than to try to offer a credible course in market milk, cheese making, butter making, ice cream, and carry on research without a college creamery.*
2. *Equip our own laboratories and purchase raw material from local plants when needed;*

a. This would necessitate equipping our laboratories with the same amount of equipment that we have at the present time and since our greatest problem is not the wearing out of equipment but the equipment becoming out of date, the cost of equipment would be practically the same as it is at the present time.

b. Would have difficulty in securing the amount and quality of raw dairy products for laboratory use when needed. We would very likely be unable to secure information on the original source of the raw material as we are under our present condition, which is valuable in teaching students.

c. Since no regular market could be established under such intermittent methods of manufacture, the product manufactured would bring nothing or a very small per cent [sic] of its original value.

d. Our teaching budget for supplies, equipment, and repairs would have to be increased considerably in order to furnish the same instruction we offer at present.

We have not mentioned all the objectives of the two methods of laboratory instruction, but we believe the above is sufficient to show the disadvantages of such methods. Dairy laboratory instruction is expensive because of the highly specialized nature of [the] dairy manufacturing industry and it seems that an institution is morally responsible for lessening the expense if it can be done.

If the Purdue University Creamery was [sic] to discontinue and we attempted to maintain our present standards of instruction and research, the cost would be considerably more than it is at the present time. As an example, the Creamery in the last four years has spent over $87,000.00 for equipment alone from Creamery Funds. This does not take into consideration the numerous supplies, repairs, and labor which the University receives at the expense of the creamery.

The Purdue University Creamery also furnishes practical training for young men in the dairy industry for it has been the policy of the Department to employ when possible men in the creamery who have had either a short course or a four-year course in dairying, or students specializing in dairying in the University.

We have 7 Federal projects and 3 State projects in which we would be considerably handicapped by increased cost, time and convenience if it were not for the Creamery, even though the material furnished by the Creamery for instructional and research work is charged to the corresponding funds.

Other Advantages

The Purdue University Creamery has furnished an excellent market for the milk produced on the Dairy Farm, considerably above the local market for milk which has been very helpful in furnishing funds for the Purdue Dairy Farm. This last year, 1957-58, the Creamery paid for milk delivered from the farm $32,314.70 for a total of 713,508 lbs. of milk, and paid the Dairy Farm an average of $.38 per 100 lbs. more than local market prices, or $2,711.32 over the local market for the year's milk supply.

The Purdue University Creamery also furnishes an excellent market for eggs produced on the Poultry Farm and during 1957-58 returned to the Poultry Department $9,651.31 for 19,936 doz. eggs.

The Creamery pays for all of the light and power used in Smith Hall and 1/2 of the water that is used. This last year the light and power bill paid from the Purdue Creamery funds was $3,744.30 and the Creamery share of the water bill was $667.32.

The Purdue University Creamery should be operated for the following purposes:

1. *To give laboratory instructions to students in the Dairy Industry*
2. *To furnish equipment and material for research and instructional work*
3. *To provide practical training for young men entering the Dairy Industry*
4. *To provide funds from the sale of products for equipment, material and laboratory space for instructional and research work*

> *If the Purdue University Creamery is not doing this we have no excuse for operation. To do the above as it should be done we must not only have modern up-to-date equipment in our creamery but we must have a modern up-to-date building convenient to our instructional and research workers.*[1]

In the second paragraph, the frustration of Professor Gregory can be felt. His patience had worn thin over the eight years since the Smith Hall additions were made. His complaints were about what he felt were certain promises which were made but were yet to be acted upon, including the construction of a new creamery.

On April 1, 1959, Gregory again sent a letter to the same three gentlemen, Butz, Volk, and Freeman. He sent them the results of a survey he had conducted with nineteen other agricultural colleges with dairy plants around the United States. This survey asked about products these colleges manufactured, the type of building used, product packaging, cheeses manufactured, and profits and how they were used.[2] With this letter, Gregory was again trying to place the Purdue Creamery in a positive light and keep the possibility of a new creamery at the forefront for the Purdue Agricultural Administration.

Gregory's effort apparently paid dividends as, on June 3, 1959, Butz sent a letter to Gregory telling him to prepare estimates on equipment costs for a new creamery.[3]

The 1960s began with the continued development of the plans to build the new creamery, but ultimately ended with the closing of the existing creamery. A proposal was made by Purdue University President Frederick Hovde to the Indiana Dairy Manufacturers Association seeking their approval for construction and operation of a new building dedicated to dairy products manufacturing. President Hovde did not want to upset any of the other dairy manufacturers and once again have the university accused of unfair competition, similarly to what occurred in 1930. Hovde's proposal was approved by the organization. The Purdue University Trustees proceeded to approve the building of a new creamery without any dissension, and the Walter Schooler Associates architectural firm was hired.[4]

As early as November 23, 1960, a decision to construct a new creamery had been made. On this date, a letter from Dr. Fred N. Andrews, the Dairy Department head, was sent to nine people at the university. It read in part: "There will be a meeting to discuss the policy matters for the operation of the new creamery at 9:00 a.m., Thursday, December 1, 1960, in Dean Freeman's office." This letter went to Dean Earl Butz, Ag Experiment Station Director N. J. Volk, University Vice-President J. C. Smalley, and six other recipients.[5]

In an effort to learn more about creamery design and sales opportunities, Animal Science Department head Andrews visited Penn State University's creamery and salesroom. He went away from that visit obviously impressed. In a letter dated December 23, 1960, to Butz, Freeman, and Volk, Andrews wrote that Penn State was also selling honey, apples, and eggs from university farms in the salesroom to supplement the dairy products sold. Andrews was told that, on some weekends, Penn State had sold as much as 1,000 pounds of Cheddar cheese, with 40,000 pounds continually stored and in the aging process. It was also said that as many as 4,000 ice cream cones could be sold on a Sunday afternoon.

The Penn State Creamery was delivering 350 five-gallon cans of milk to the dorms daily. The male students in these dormitories were averaging eleven quarts of milk each week with the women averaging about nine quarts. They were also consuming about 300 gallons of soft serve ice cream each week as well.[6]

In late September 1961, Dr. Andrews sent a letter to Professor Rhodes, Dr. Babel, and B. J. Liska. According to the letter, the planning of the new creamery was well underway and pretty much set in stone. The letter reads:

> *Dr. Liska is to be working with the architect on new creamery plans during the school year. It is important that he be fully aware of all creamery, research and teaching functions if he is to be valuable to the department in this assignment. I hope you will have frequent meetings to discuss mutual problems.*[7]

In July 1962, the Hertel, Johnson, Eipper, and Stopa firm of Chicago was hired to consult with Walter Schooler Associates on the interior design and needed equipment for the creamery. The Chicago firm was a specialist in the dairy building construction design industry. The hiring was approved by the Board of Trustees because Walter Schooler Associates needed expert assistance at that point in time.[8]

12
Casualties Occur

On Sunday, September 28, 1962, the plans for a new creamery began to come undone. A story found on the front page of *The Indianapolis Star* entitled, "Purdue's Research Creamery Doesn't Make the Butter It Sells," was the start of difficult times for the future Purdue University Creamery. Many believe it resulted in the closing of the old creamery less than seven years later. The article was written by Edward H. Frank and reads as follows:

> *The purpose of the lucrative and tax-exempt Purdue University creamery, which produces everything from egg nog to Swiss cheese, is "research and education," Purdue officials claim.*
>
> *But Purdue educators have trouble explaining why the creamery sells "Purdue University butter" in supermarkets and on home delivery routes at West Lafayette when the butter isn't even made in the Purdue creamery.*
>
> *The butter is produced and packed in Purdue cartons by the Midwest Producers Creamery of South Bend and then shipped to West Lafayette where the university competes with tax-paying dairies by selling the butter commercially.*
>
> *So research in packaging butter isn't a justification for the sale of the butter by the Purdue Creamery.*

Valley Lea is the brand name of the Midwest Producers' butter.

But at West Lafayette, thousands of persons probably think they are buying Purdue Creamery butter because it is packed in flashy gold and black cartons which are marked "Sweet Cream Purdue University Creamery Butter."

At the very front at the bottom of the carton, is the word "distributors."

Last year the Purdue creamery grossed $515,631 selling milk, coffee cream, cottage cheese, Swiss cheese, chocolate milk, eggs, buttermilk, non-homogenized milk, sour cream, butter, egg nog, about a dozen flavors of ice cream and a variety of sherbets.

The creamery showed a net profit at $40,000.

Earl. L. Butz, Dean of the Purdue School of Agriculture, said he questioned the retail sale of butter nine years ago when the Purdue creamery stopped producing butter because of economic conditions in the dairy industry.

However, Butz pointed out that families buying the Purdue creamery products from Purdue milkmen expect to be able to buy butter.

Purdue butter is sold in the supermarkets, Butz said, because it provides a complete dairy line along with cottage cheese and Swiss cheese which are sold in the supermarkets.

"When we stopped producing butter in 1953, we had to buy sufficient amounts in order to get low prices for the butter we needed in the dormitories," Butz said.

This results in an excess which the university sells commercially, Butz said.

Last year, Butz estimated that Purdue sold 15,000 to 20,000 pounds of butter in West Lafayette supermarkets and home delivery routes.

"We're phasing out of the milk routes," Butz said, reporting that only two routes now are in operation.

Butz said that 80 per cent [sic] of all gross sales by the creamery are to the Purdue dormitories and the Union Building. The other 20 per cent [sic] is divided between the supermarkets and home delivery.

Last year, the creamery sold 400,000 gallons of milk, 26,000 gallons of ice cream, 5,000 gallons of sherbets, 101,000 pounds of cottage cheese, 69,000 pounds of butter and 17,000 pounds of Swiss cheese.

The entire need for the huge operation at Purdue is questioned by some leaders in the dairy industry.

"If it were up to me, I'd close it up," said Greyble L. McFarland Jr., President of Golden Guernsey Farms, Inc. Indianapolis.

"Purdue should concentrate on research in the dairy industry rather than worrying about running a big business," McFarland said.

McFarland and Robert A. Scott, manager of the Borden's plant in Indianapolis, pointed out that the need for dairy technologists has dropped as the few remaining dairies become more and more automated.

"As far as I'm concerned," Scott said, "I need engineers more than I need someone who studied dairy technology."

Figures show that only 28 students now are studying dairy production and dairy manufacturing at Purdue. Only 26 were enrolled last year and 21 the year before.

And these students do not even work in the creamery, according to F. N. Andrews, head of Purdue's Department of Animal Sciences.

The students do some laboratory and observation work in the creamery. But the actual work is done by salaried employees.

Andrews stated that research is the chief purpose of the creamery. And in order to conduct intense research, vast supplies of milk and milk products must be used. This is the reason Purdue entered the commercial dairy business.

About 25 per cent [sic] of the raw milk supplied the dairy comes from the Purdue dairy herds and the other 75 per cent [sic] comes from 15 privately owned dairy farms.

Andrews said research in the creamery has brought many advances to the dairy industry.

Purdue Swiss cheese has become nationally famous.

Dr. F. J. Babel developed a dressing for cottage cheese at Purdue which lengthened the keeping qualities of the cottage cheese several days.

"This development is worth millions to the dairy industry each year." Andrews said.

W. K. Holm, executive secretary of the Indiana Dairy Producers' Association, said his organization feels the Purdue creamery is worthwhile.

"We would only be concerned if Purdue decided to expand its retail sales into Lafayette proper," Holm said.

With the profits earned each year from creamery operations, Purdue plans on building a new plant.

If this money wasn't available, Purdue would have to ask the Indiana General Assembly for funds to build a new plant, said Lytle J. Freehafer, Purdue vice president and treasurer.

The $40,000 profit the tax exempt creamery earned last year represented a net profit of 8 per cent [sic] of sales.

Most tax paying dairies are lucky to net 2 per cent [sic] today according to McFarland.

"To build a new creamery plant at Purdue would be as foolish as it is to operate the present one," McFarland said. [1]

Purdue's response to this article was quick and somewhat forceful, although ineffective in the long run. During the October 24, 1962, Board of Trustees meeting, President Hovde told the board that *The Indianapolis Star* reporters had lost their assignments in regard to reporting on the university's activities. Hovde still felt that the issues involving the creamery were manageable and under control because Purdue had the backing of the Indiana Dairy Producers Association. Most dairy product processors in Indiana belonged to this organization. Despite Hovde's confidence, officials at Purdue still felt the sting of this article as evidenced by the fact that the secretary who wrote the minutes of the meeting labeled it as "Press Attack on University Activities."[2]

Dean Butz's concerns about manufacturing butter away from Purdue's campus were noted not only in the newspaper article but also in a letter he sent to Assistant Dean of Agriculture Verne C. Freeman on October 10, 1960. This was nearly two years prior to the article printed in *The Indianapolis Star.* The letter from Dusty Rhodes that is mentioned in this letter by Butz has not been located.

I am returning here with the communication to you from Dusty Rhodes about butter sales by the creamery.

It seems to me that as long as we have the retail routes, we will need to have butter for patrons on those routes. This probably is also true of the milk patrons.

Like you, I seriously question the advisability of continuing butter distribution on the wholesale route, particularly across the river, unless for some reason this is needed to keep our outlets for cottage cheese. However, I can't see any reason why this would be true.

Before making any final decision, you probably ought to assess the impact on creamery earnings, if any, of discontinuing butter distribution across the river through stores.[3]

Purdue eliminated the production of butter in 1953, so it is confusing why these concerns were not raised until 1960. Additionally, what changed that made it a problem to sell butter east of the river into Lafayette?

Another mystery from 1958 may provide some of this information. On the outside of a manila folder, in the upper-right corner, was a green-inked, handwritten note from Assistant Dean Freeman to Dean Butz:

Dear E. L. Butz,

You mentioned the creamery situation—here are the reports by months—to date.

This was followed in the lower-right corner of the folder by:

Note: They have come to me this year for the <u>first</u> time—in other words I have no previous years [sic] data for monthly comparison—except as may be shown in the reports. V.

Inside the folder were monthly financial reports for the preceding twelve months. Each month's report was twelve pages in length with detailed incomes and expenses. Production totals for each product sold were also included. There was no other reference to "the creamery situation."[4]

Listed below are the butter purchases from outside sources by the Purdue Creamery starting in 1928. It is unclear whether the creamery was purchasing butter prior to 1928. When the amount of money spent on butter prior to 1962 to 1963 and later is compared, it is clear that a very large decrease in dollars spent occurred. This decrease in butter purchased was obviously the response by Purdue to *The Indianapolis* Star article. Unfortunately, it is not known how many pounds of butter were purchased, only the dollars spent on the purchasing of the butter. These figures were found in the *Purdue University Bulletin-Financial Reports* for each respective fiscal year. These reports are located in the Purdue University Library Archives.

VALUE OF BUTTER PURCHASED BY PURDUE CREAMERY			
Year	**Value ($)**	**Year**	**Value ($)**
1928	$4,275.31	1949	Not listed
1929	$11,897.33	1950	Not listed
1930	$10,987.69	1951	$8,739.09
1931	$11,738.98	1952	$28,150.20
1932	$2,573.22	1953	$51,251.04
1933	Not listed	1954	$49,772.16
1934	Not listed	1955	$41,864.26
1935	$1,665.95	1956	$48,960.89
1936	$8,666.72	1957	$52,386.07
1937	$5,150.00	1958	$40,565.23
1938	$5,546.56	1959	$38,480.80
1939	$1,243.57	1960	$39,580.86
1940	$4,295.73	1961	$42,627.40
1941	$1,584.70	1962	$44,966.89
1942	Not listed	1963	$19,972.26
1943	Not listed	1964	$10,468.77
1944	Not listed	1965	$7,967.58
1945	Not listed	1966	$7,527.88
1946	Not listed	1967	$6,143.89
1947	Not listed	1968	$5,582.11
1948	Not listed	1969	$1,008.89

Below profit and loss statements for an eight-year period, from 1954 to 1963, are listed. This statement was not a finalized report; adjustments apparently still had to be made. Because of this, when 1956 through 1963 are compared, the profits on the two charts do not match. Also listed in the first table are the quantities of milk, ice cream, cottage cheese, and butter sold during that same timeframe. This information can be found in the *Purdue University Bulletin-Financial Report* for each year listed. These reports are located in the Purdue University Library Archives. Any information used in the written text has the year, volume, number, and page cited accordingly.

	YEAR							
	1956	**1957**	**1958**	**1959**	**1960**	**1961**	**1962**	**1963**
Profit Ending June 30th of Each Year	$37,863	$36,422	$34,427	$41,735	$30,171	$39,830	$53,579	$36,352
Milk Sold, Fluid Gallons	312,024	336,786	346,930	339,976	357,705	380,019	400,182	385,072
Ice Cream Sold, Gallons	37,462	36,424	36,858	36,268	36,829	40,120	43,242	41,920
Cottage Cheese Sold, Pounds	68,370	71,750	66,259	74,954	93,166	104,909	101,742	72,625
Butter Sold, Pounds	86,716	83,185	64,982	62,242	65,963	66,678	69,594	33,540

PROFITS OR LOSSES OF PURDUE CREAMERY 1926–1969			
Year	Value ($)	Year	Value ($)
1926	$3,290.32	1948	$16,581.47
1927	$299.28	1949	$71,847.47
1928	-$2,243.24	1950	$59,534.83
1929	-$1,126.71	1951	$14,605.29
1930	-$2,140.43	1952	$3,210.65
1931	$4,307.40	1953	$8,713.22
1932	$1,869.80	1954	$7,197.27
1933	$180.57	1955	$9,753.71
1934	$1,182.52	1956	$40,700.79
1935	-$103.65	1957	$38,052.66
1936	$1,872.48	1958	$42,251.57
1937	$8,243.03	1959	$47,245.89
1938	$9,429.68	1960	$39,580.86
1939	$23,257.56	1961	$40,627.40
1940	$14,233.29	1962	$64,279.18
1941	$9,259.02	1963	$48,591.19
1942	$1,267.77	1964	$46,947.71
1943	$9,930.18	1965	$3,609.19
1944	$11,031.83	1966	$53,111.44
1945	$11,555.80	1967	$21,735.37
1946	$23,313.89	1968	$19,436.77
1947	$49,343.25	1969	-$1,999.78

Information prior to 1926 is not known as a profit and loss statement, but rather as receipts and expenditures. These do not truly indicate the profits or losses for that year. The following can also be found in the *Purdue University Bulletin-Financial Reports.*

1924 — Receipts of $129,919.31 Expenditures of $122,072.24
1925 — Receipts of $134,986.27 Expenditures of $136,085.32
1926 — Receipts of $138,276.59 Expenditures of $139,456.53

When looking at the actual paperwork with monthly figures of profits or losses for the above years, every June, July, and August were financial losses. This happened because students were gone for the summer months, so consumption of milk and milk products plummeted.[5]

There is evidence of the Purdue Creamery also selling eggs as early as 1951, which continued until 1969 when the creamery closed. As many as 21,000 dozen per year were purchased from the Purdue

University Poultry Farm for resale.[6] Unlike with butter, no criticism of the creamery's purchase of a large number of eggs has been found. The value of eggs purchased each year can be found in the following table. When viewing the years from 1951 to the closing of the creamery, similar trends in purchases can be seen for eggs as with butter. The total number of dozens of eggs purchased is not available in these reports, only the total yearly monetary value. The figures below were found in the *Purdue University Bulletin-Financial Report* for each respective fiscal year. These reports are located in the Purdue University Archives.

VALUE OF EGGS PURCHASED	
YEAR	VALUE ($)
1951	$5,376.81
1952	$6,492.36
1953	$7,354.03
1954	$9,127.92
1955	$7,781.84
1956	$9,337.13
1957	$8,807.56
1958	$9,651.31
1959	$7,976.90
1960	$6,308.58
1961	$6,615.71
1962	$5,097.95
1963	$5,340.57
1964	$4,154.87
1965	$3,214.97
1966	$2,061.01
1967	$2,150.63
1968	$1,050.21
1969	$1,008.59

In June 1963, Dusty Rhodes began writing purchase orders for the upcoming fiscal year starting July 1, 1963. Even though there are no values to compare these purchase orders to, the quantity of supplies ordered seems quite large. Included in the orders were 800 dozen large eggs and 500 dozen medium eggs per month, totaling 15,600 dozen for the year. Eggs were used in ice cream production and retail sales. Also ordered were 25,000 ice cream Safe-T Cups for individual ice cream sales, 50,000 wooden ice cream spoons, and 12,000 one-quart ice cream cartons.

Flavorings ordered for ice cream and sherbet were chocolate, strawberry, cherry, pineapple, lemon, orange, lime, orange peel, and lemon custard. The concentrates for orange drink, grape drink, and fruit punch were also ordered. There were also orders for cardboard pour-spout type containers. Five thousand quart containers and ten thousand one-half pint containers for each of the three flavors were ordered from the Weyerhaeuser Company of Seymour, Indiana.

Gold and black cardboard milk cartons were purchased from the Weyerhaeuser Company. Ordered were 17,000 quart cartons, 120,000 one-half pint cartons 7,000 one-color cartons for regular pasteurized milk, 5,000 quarts and 17,000 one-half pints of chocolate milk, and 5,000 quarts and 14,000 one-half pints of low-fat milk.

Five thousand quart and 7,000 half pint buttermilk cartons were needed, as were 5,000 quart containers for skim milk, 5,000 quart and 5,000 pints of half-and-half containers. Five thousand quart and 5,000 half pint cardboard whipping cream containers were ordered.

Rhodes needed 10,000 gold-and-black five-pound cottage cheese containers from General Dairy Supply of Indianapolis. Half of this order was supposed to arrive at the creamery on February 1, 1964, and the other half on July, 1, 1964.

In contrast to all of the paper cartons ordered for the dairy products, Rhodes ordered 60 gross or 8,640 square, tall half-pint glass milk bottles without any lettering on them. These bottles had tops designed for 48 mm tops or openings for Stanseal caps. These bottles were priced at $6.00 per gross for a total of $360. Two hundred wire milk cases were ordered to carry 16 cardboard cartons for quarts or 44 half-pint cartons.

Collection of Purdue Creamery items.
(Courtesy of John Cleland)

Sixty thousand pounds of block ice would be needed for the upcoming year. The Physicians Clinical Laboratory of Lafayette was paid $210 for analyses of milk and milk products. These analyses were independent and unbiased checks of bacterial counts and sanitation control.

White uniforms were furnished to the creamery and salesroom personnel. The uniforms ordered may indicate the number of workers employed at the creamery. A total of eighteen dresses were ordered, six were size 32, six were size 34, and twelve were size 36, which indicates that at least four different women were employed, most likely in the retail salesroom.

For male workers, twenty-four pairs of white pants were ordered, twelve were size 34 x 30, six were 32 x 30, and six were 30 x 32. Twenty-four white shirts were ordered, twelve were neck size 15, and twelve were neck size 15 1/2, which indicates that four men were probably employed in the creamery itself.[7]

Following the September 1962 article published in *The Indianapolis Star*, creamery production did not immediately slow down as judged by the electricity used by the facility. The power use actually increased by 74,159 kilowatt hours from July 1, 1963, to June 30, 1964, compared to the preceding calendar year.[8] Power consumption did not start to decline until the 1964–1965 budget year, when it declined by more than 10,000 kilowatt hours.[9] The criticism levied by the article did, however, force the creamery to eliminate the remaining two home delivery routes. The minutes of the May 31, 1963, Board of Trustees meeting does indicate that a new creamery was still listed as a construction project.[10] The construction project was again noted in a letter from University Vice President and Treasurer, Lytle J. Freehafer, dated June 18, 1963.[11] The proposed creamery planned to have a drive-in area for retail sales. This drive-in retail area was planned as a remedy for customers still desiring Purdue Creamery products after the discontinuation of the home delivery routes.[12]

Although nothing of significance about the creamery is found in minutes from the Board of Trustees meetings from May 1963 to January 1968, plenty was happening within the Purdue administration. A June 18, 1963, memorandum from Dean of Agriculture Earl Butz in regard to "New Creamery Construction," details his goal to cease home retail route sales by September 1, 1963. This led to discussions about terminating the sales of Purdue Creamery products on home delivery routes, which were completed by the middle of July 1963. These discussions were held in the office of Dean V. C. Freeman. In attendance with Freeman were Dr. Krider, Animal Science Department head, V. D. Rhodes, superintendent of the creamery, and Dr. Ralph Erb and J. O. Young, both from the Animal Science Department.

During the first half of 1963, four retail routes were still operating and were split between two drivers, each driving trucks that were seven years old. A total of 450 customers were on these routes. Selling these routes to a local dairy was discussed first, but that idea was soon dismissed because it had too many negatives, including criticism from the losing dairies and a fear that the winning bidder's products might not meet the high quality and standard of Purdue dairy products. Freeman and the others in the meeting also considered that the other dairies would try to lure the customers away from the new buyer, which again might produce ill will and bad feelings toward Purdue.

As the meeting continued, the attendees decided to give the customers a two-week notice prior to ending the routes. The two drivers would also be given this two-week period of time. The committee members knew that one of the drivers wanted to go into business for himself. The second driver would be brought back to work inside the creamery. He was also free to seek employment as a route driver for one of the other four dairies still operating in the Lafayette area. These four local dairies are believed to have been Borden's, Grandview Dairy, Rose Glen Dairy, and Sunshine Dairy. The two leftover delivery trucks would be sold as surplus through the Purdue Accounting Department.[13]

These conversations also included the creamery's reduced profits, which were estimated to be about $15,000 annually, as a result of these routes being removed. Although at that time, the routes did not earn $15,000, some unreducible fixed costs that had been charged to the routes would have to be charged to some other area of the creamery. Removing these two routes would ultimately reduce total profit.[14]

After shutting down these home delivery routes, the delivery men were missed, which is expressed in this short article from the *Lafayette Journal & Courier,* in a quote from Fern Martin:

> *I have fond memories of the Purdue Creamery operations, and not just the fabulous ice cream cones at Smith Hall. When I was a little girl, our milk was delivered to our home from the Purdue Dairy. I can remember that my mother always specified milk from "dairy herd" as opposed to "creamery milk." Purdue retailed and delivered a full line of dairy products, including cream, whipping cream, buttermilk, butter and cottage cheese. And of course, that yummy ice cream!*[15]

The term "dairy herd" in the preceding paragraph refers to milk that was produced specifically from the cows located on the Purdue Dairy Farm as opposed to milk purchased from local dairy farmers. This milk was processed and bottled separately from the purchased milk. Milk bottle caps can be found with this "dairy herd" designation.

Milk bottle cap used for Purdue dairy herd milk.
(Courtesy of John Cleland)

Another major problem with discontinuing home delivery route sales was the standing of the creamery under the Federal Milk Marketing Order. Under Federal Law established in 1937, the government developed a pricing formula for each of the four classes of milk: Class I is fluid or beverage milk; Class II is milk used for soft manufactured products such as yogurt, cottage cheese, and cream; Class III is for hard cheeses; and Class IV is for butter and dry milk.

When Class I usage declines and a surplus occurs, normally the excess milk is put into one of the other classes that has a longer shelf life. To receive a special pricing discount in the 1960s, fifty percent or more of the milk brought to the processor had to be used for Class I or fluid milk purposes. This discount to Purdue was worth as much as $5,000 to $6,000 annually. When students were away from campus, their absence put a real strain on the Purdue Creamery to meet this requirement. During those times, the retail sales routes provided a small cushion to help meet this federal requirement.

To address this issue after discontinuing the sales routes, the committee thought that one possibility was to ask for a special exemption from the Federal Milk Marketing Order to allow them to fall below the fifty percent level. This would require special hearings and take up to twelve months to complete.[16]

The reduced profit was apparently being felt by May 1964. F. D. Odell, the representative of the comptroller for the creamery and Animal Science Department, sent a letter to twelve Animal Science administrators, School of Agriculture administrators, and Purdue University administrators outlining a list of events that had taken place or would take place in the near future to reduce creamery labor costs. Besides those mentioned above, other actions mentioned in the letter were as follows: (1) One wholesale route was eliminated and the route driver was let go in November 1962; (2) the assistant creamery superintendent was transferred to another position within the university on January 1, 1964; (3) the retail and inventory clerk position was removed, effective April 15, 1964; (4) one clerical position was transferred to another university position on July 1, 1964; and (5) one plant person was retired by the end of 1964. Mr. Odell also wrote that vending machines for milk would be placed around campus, and sales of milk cartons would begin on July 1, 1964.[17]

Originally, one of the purposes of the creamery was for training and education of students in the production of dairy products. Although class sizes were small, trained dairymen were still being sought in the marketplace into the mid-1960s. In 1964, only three students graduated with a Dairy Manufacturing degree. Myron Bozell remembers manufacturing classes with less than a handful of students in them. In addition, most of the classes were offered only every other year. This meant students and their academic advisors had to thoughtfully prepare their course schedules so that a class wasn't missed and thus require students to enroll in extra semesters to graduate.[18]

The low numbers of trained graduates were felt by the dairy plants of Indiana. A half page ad in the *Purdue Agriculturist* in March 1965, paid for by the Indiana Dairy Products Association, advocated for student opportunities for summer and full-time employment. It also suggested students investigate scholarship availability and curriculum requirements.

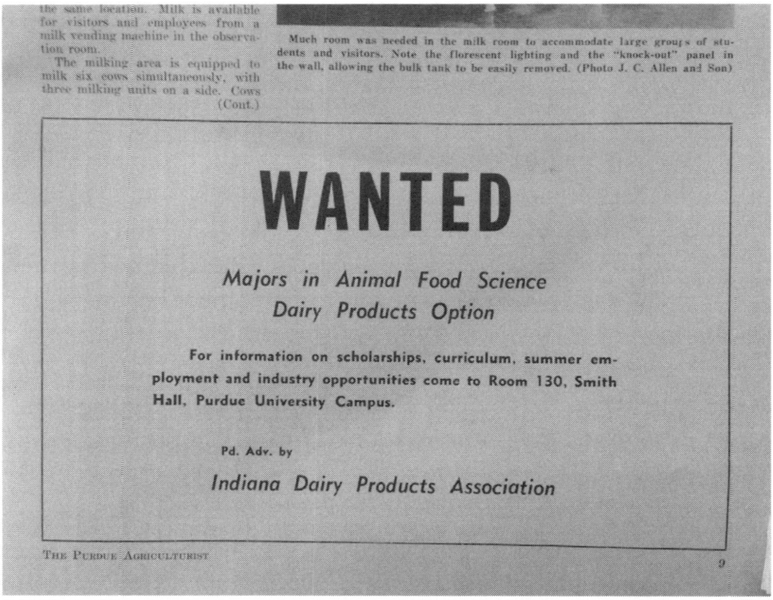

the same location. Milk is available for visitors and employees from a milk vending machine in the observation room.

The milking area is equipped to milk six cows simultaneously, with three milking units on a side. Cows (Cont.)

Much room was needed in the milk room to accommodate large groups of students and visitors. Note the florescent lighting and the "knock-out" panel in the wall, allowing the bulk tank to be easily removed. (Photo J. C. Allen and Son)

WANTED

Majors in Animal Food Science
Dairy Products Option

For information on scholarships, curriculum, summer employment and industry opportunities come to Room 130, Smith Hall, Purdue University Campus.

Pd. Adv. by

Indiana Dairy Products Association

THE PURDUE AGRICULTURIST 9

Advertisement for more Dairy Manufacturing graduates.
(Courtesy of Purdue University Archives and Special Collections, Purdue University Libraries)

In the same issue of the *Purdue Agriculturist,* an article expanded on the sentiment of the advertisement:

> *The dairy industry is a large and fast growing industry. It includes areas of opportunity in teaching, research, administration, dairy engineering, laboratory control, inspection and grading, and merchandising. A skilled dairyman may manage, operate, or own a dairy plant. Dairy manufacturing includes many facets—ice cream, cheese, milk processing, condensing and dry milk manufacturing, dairy bars, retail dairy products stores and multi-products plants.*
>
> *Because it is a food-producing industry, dairy manufacturing is affected very little by changes in business cycles. Even the depression of the 1930's [sic] had very little effect on the dairy industry. The demand for milk is year round and the supply is continuous giving dairy workers a secure job as well as a good paying one. The starting salary for graduates in dairy manufacturing is $6,000 to $7,500 per year. After a few years of experience, salaries are increased to $8,000 to $10,000, or more.*

Because sanitary conditions are required dairy plants are clean, modern, and efficient, therefore working in them is pleasant.

Opportunities for advancement are excellent because of the rapid growth of the dairy industry in all areas. The growth of population along with the greater appreciation for the importance of dairy products in the diet leads to ever increasing sales. More milk, in all of its forms, is consumed per person than any other single food. Sales are over ten billion dollars a year and over thirty five [sic] thousand plants are in operation in the United States today. Constant improvements in marketing and processing are creating larger and better markets everyday [sic]. To illustrate this, newly designed plants have been developed with operations programmed entirely by computers and which hire only trained dairy personnel, all of which reduces the overall cost of the finished dairy product. Today laboratory discoveries will result in common products in the markets of tomorrow.

Due to the rapid growth of the industry and the development of new forms of dairy products, the industry is faced with a shortage of trained personnel. Many dairy companies have recognized the need for more trained young people in the field of dairy manufacturing and are supporting scholarships which are available to those studying dairy technology.

The Chicago Dairy Technology Society offers a $2,000 scholarship each year to a qualified freshman student who wishes to study Dairy Science. The Indiana Dairy Products Association provides $300 each year for the Otto F. Hunziker scholarship for a junior who is studying Dairy Technology and $300 for the H. W. Gregory scholarship for a senior who is studying Dairy Technology.

The Institute of Food Technologists offers 30 scholarships of $300 each to freshmen and sophomores who are studying Food Science and Technology.

In addition to these scholarships, many others are available to qualified students through the Purdue Scholarship Office.[19]

In the late 1950s and early 1960s, dairy manufacturers began pointing out inadequacies in the training and education of students by universities throughout the United States. Job descriptions within dairy plants began to change rapidly as new technology was introduced. These manufacturers began to suggest and request changes within the college curricula to meet their manufacturing needs. One example was found in a letter sent by Tom Maynard to Dr. Fred Andrews. Mr. Maynard was the Merchandising Director for the Quality Chekd Dairies Manufacturers Association of LaGrange, Illinois.

Maynard suggested putting less emphasis on actual training and use of equipment. He thought science and technology was changing too rapidly to spend large amounts of time on practices soon be outdated. He also suggested that colleges give more training in economics, management, and sales.

Dr. Andrews shared this letter with Dean Butz. Butz excitedly responded by writing a letter to Maynard that was just about in total agreement with Maynard's sentiments. Butz pointed out that Purdue had already made some revisions and that more would soon be implemented. He wrote, "To most of your letter I have to say a hardy amen."[20]

Shortly after these pleas were made for trained dairy technologists, new technology came to the Purdue Dairy Department and the creamery. A machine called the Infra-Red Milk Analyzer, or I.R.M.A, began to replace the Babcock test for analysis of cream content within milk and dairy products. The Babcock test had been the standard test for butterfat percentage since the late 1800s and had proved to be very accurate. However, the Babcock process was time-consuming. It took approximately three to four minutes to test each sample using the Babcock method, but just over thirty seconds per sample with the I.R.M.A.

Purdue tested the newly designed Danish-engineered machine in studies done in late 1966 and 1967. The machine proved to be quite accurate and reliable as long as certain factors were taken into account. First, milk proteins affected the accuracy and had to be solubilized. Second, if they were highly acidic, milk preservatives had to first be neutralized. And third, homogenization of milk also caused false readings.[21]

The author used one of these machines extensively while employed part-time as an assistant Creamery License inspector during late 1973 and 1974. They were easy to use, easy to clean, safe, and fast. They were checked and calibrated on a regular basis against the Babcock test by using the same samples of milk. An additional benefit of the I.R.M.A. was that the Babcock test required the use of highly concentrated sulfuric acid, so the potential of a spill or inadvertent contact was always present.

At the January 16, 1968, Board of Trustees meeting, creamery product prices were changed. This change was reported January 16, 1968, but actually took place December 1, 1967. These changes were for deliveries involving the campus routes and for sales through the retail salesroom. The product prices for both campus routes and retail room sales are shown below.[22]

CAMPUS ROUTE RETAIL MILK PRICES				
	Gallon	**1/2 Gallon**	**Quart**	**1/2 Pint**
Homogenized	-	-	$0.23	$0.0625
Chocolate Milk	-	-	$0.25	$0.0675
Cultured Buttermilk	-	-	$0.23	$0.0625
Half & Half	$2.82	$1.42	$0.73 (or $0.37/pint)	-
Coffee Cream	$3.74	$1.88	$0.96	$0.26
Whipping Cream	$5.95	$3.05	$1.55	$0.42
Cultured Half & Half	$2.86	$1.45	$0.75	-
Cultured Coffee Cream	$3.80	$1.92	$0.98	-

CAMPUS ROUTE ICE CREAM PRICES					
	5 Gallon	**2 1/2 Gallon**	**I Gallon**	**1/2 Gallon**	**Sl. Bricks**
Vanilla	$1.45	$1.48	$1.51	$1.54	$0.48
Chocolate	$1.55	$1.58	$1.61	$1.64	$0.48
Other Flavors	$1.55	$1.58	$1.61	$1.64	$0.48
Sherbets	$1.45	$1.48	$1.51	$1.54	-

These ice cream prices are on a per gallon basis. In other words, a customer did not pay $1.45 for five gallons of vanilla ice cream and $1.54 for just a 1/2 gallon. In contrast, bricks of ice cream were not priced on the per gallon scale.

Because the smallest unit of American currency is a penny, it is strange to think that 1/2 pint of milk sold for slightly more than six cents per carton. This can be explained by remembering that, on a route with fraternities, sororities, co-op houses, and dormitories, these institutions might be purchasing dozens or even hundreds of half pints at a time. What remains a mystery is the fraction of a cent pricing on chocolate milk as printed on the retail room price list. It is unknown if this was a typing error or not.

RETAIL ROOM DAIRY PRICES					
	Gallon	1/2 Gallon	Quart	Pint	1/2 Pint
2% Low Fat Milk	-	-	$0.25	-	$0.07
Creamery Milk	-	-	$0.26	-	$0.07
Homogenized Milk	$0.90	$0.48	$0.26	-	$0.07
Chocolate Milk	-	-	$0.28	-	$0.075
Cultured Buttermilk	$0.90	-	$0.26	-	$0.07
Skim Milk	$0.72	-	$0.24	-	-
Half and Half	$2.94	$1.52	$0.78	$0.40	-
Coffee Cream	$4.20	$2.20	$1.15	-	$0.30
Whipping Cream	$6.30	-	$1.65	-	$0.45
Polygal Homogenized	$0.90	-	-	-	-
Polygal 2%	$0.86	$0.45	-	-	-
Cultured Half & Half	$3.12	$1.56	$0.78	$0.33	-

Although many dairy products changed price, ice cream sold through the Smith Hall salesroom stayed the same price. Only in November 1968 did the prices for ice cream cones change, doubling from a nickel to a dime. This was the first known price increase for ice cream in the history of the creamery at Smith Hall.[23] In fact, there were even situations when ice cream was free. Elaine Baumgardt recalled that employees of the Dairy Department and their families could get free ice cream as long as it was consumed on the premises. Elaine's husband, Bill Baumgardt (B.S. 1955, M.S. 1956), worked for Purdue. As he pursued his degree in Dairy Husbandry, Bill was also under the direction of Dr. J. W. Wilbur. Nearly every afternoon when Bill and Elaine's two-year-old daughter, Pam, woke up from her nap, Mrs. Baumgardt put Pam in a stroller. She then walked from their married student court apartment, located in old World War II military barracks along Airport Road, to the salesroom in Smith Hall. There Mrs. Baumgardt received a free chocolate ice cream cone for herself and a vanilla ice cream cone for Pam.[24] In 1985, Bill Baumgardt was elected as president of the American Dairy Science Association. He is one of five people with a background linked to Purdue who have been elected to this office.

Students crowding into creamery salesroom.
(Courtesy of Purdue Agricultural Communications Department)

The price increase was not enough to fend off declining profits. Each year after 1962 (except for 1966) contributed to a downward trend in the creamery's profitability. During that time, equipment was not being replaced or modernized, but few people were ready for the shock that was about to occur.

Without prior notice to the public, Vern Freehafer, Purdue University treasurer, announced the closing of the Purdue Creamery at the April 24, 1969, Board of Trustees meeting. The board minutes read as follows:

REPORT OF DISCONTINUANCE OF PURDUE CREAMERY

Treasurer Freehafer reported that methods of processing and distributing dairy products have changed materially in the past fifteen or twenty years. Because the P.U. Creamery is technically outmoded, its equipment is ancient, and creamery operation for even one more year would require substantial capital investment, University officers and staff have decided to discontinue operation of the Creamery as of June 30, 1969. They believe that dairy products for the Purdue Memorial Union and University residences can be purchased more economically than they can be produced by the Creamery. The staff in Animal Sciences hopes to develop a pilot operation to provide basic instruction and a research program of real competence. Vice President Andrews reported that work in that area will be related to that in many other areas in connection with the Food Sciences Institute, in preparation for participation in solution of world food problems.

After discussion, the foregoing report was received with nostalgic regret for the loss of the long famous Purdue cheese and ice cream.[25]

In its article about the closing of the creamery and salesroom, the *Purdue Alumnus* called it "a campus ice cream oasis for generations of snack hungry students." This article also mentioned bits of the discussion during the Purdue Board of Trustees meeting. For example, trustee Mrs. Luke White of Covington, Indiana, asked, "Does this mean that those good ice cream cones and Swiss cheese will be casualties?"

Vice President and Treasurer Lyle J. Freehafer nodded sad assent. An unnamed administrator also commented, "Good ice cream cones have sort of become a national casualty." Freehafer thought that Purdue could purchase processed dairy products from suppliers "about as cheaply as we were producing it."[26]

In an interview during the 2014 Indiana State Fair, Mauri Williamson was outspoken about the closing of the creamery forty-five years earlier. Williamson, the former executive director for the Purdue Ag Alumni Association, said, "I absolutely loved our Agricultural Administration, but the one thing I totally disagreed with was the closing of the creamery." He went on to say that a university creamery is a great public relations piece, not only for an agricultural school, but for the entire university.[27]

While Williamson was bothered by the closing, Dairy Extension man Dr. Will Dillon was more pragmatic. His first thoughts were, "It was probably about time." He went on to say, "Equipment was worn out and very few students were being trained. It just wasn't economical or practical."[28]

On June 6, 1969, Purdue awarded Prairie Farms a contract to supply the university with most of its dairy products for the fiscal year running from July 1, 1969, to June 30, 1970. The contract was worth $311,000. A contract for ice cream was also awarded to Sealtest at a cost of $25,000.[29]

Purdue Creamery cottage cheese and ice cream boxes.
(Courtesy of John Cleland)

On the last day of the creamery, hundreds of patrons stopped by the salesroom for a final ice cream cone. Earl Butz stopped and purchased a black cherry cone. John Hicks, special assistant to President Fred Hovde, also stopped by for his last time. Also stopping for ice cream were many 4-H members (on campus attending the annual 4-H Round Up) along with faculty members, college students, former students, and local citizens.[30] On June 30, 1969, at the end of the day, Dusty Rhodes put his keys in the door and locked it for the final time. The next day, July 1, the lights to the creamery and its salesroom remained dark, and the door remained locked.[31]

The Creamery License Division soon occupied the former salesroom, filling it with butter fat testing machines and Babcock test equipment. For several years, people would stop by Room 115, the former salesroom, and ask about buying Purdue ice cream, not knowing the creamery had closed. They were merely seeking to taste that delicious ice cream one more time and bring back memories from long ago. In the rear of Smith Hall, filling the former creamery and many of its stainless steel tanks were aquaculture projects and food science endeavors.

In 1974, a wing was added to the southwest corner of Smith Hall for meat cutting and laboratory work. It was a USDA inspected slaughter house.[32] The $950,000 cost of developing this wing was paid for by leftover creamery savings funds earned prior to the closing.[33] This facility is still open and in operation today. In 1979, additional changes and modifications to the building were approved in anticipation of bringing in the department of Food Sciences.[34] In 1991, according to Thomas Schmenk, director of facilities, planning, and construction for Purdue, the old creamery was falling apart because of almost one hundred years of continual exposure to moisture and high humidity.[35] On March 21, 1992, it was announced that the old, worn-out creamery was too difficult to properly maintain and use. It was designated to be demolished soon thereafter.[36] This empty space quickly turned into additional parking.

Smith Hall, 2016.
(Courtesy of John Cleland)

The Animal Science dairy staff left Smith Hall for other buildings when Food Science moved in. At the time of this writing (2017), Smith Hall is the home of the Purdue Entomology Department.

The old creamery salesroom is full of large electronic machines used as incubators and coolers for raising and hatching insects. The salesroom doors now have warning signs about potentially toxic chemicals that are inside instead of an advertising sheet with the ice cream flavor of the day.

On the south side of the main east-west hallway hangs a portrait of William C. Smith. This portrait was donated to Purdue University by several of his nieces and nephews in 1921.[37] Their names can still be found on the engraved plate attached to the original frame. The black-and-white photograph is said to be "an excellent likeness." Next to this is a photograph of Smith Hall when it first opened in 1913. Today countless students and professors walk past this photograph and never bother to look at it. Most never appreciate the generous gift from Smith or understand the rich history and the dedication to Indiana and American agriculture that Smith Hall, the Purdue Creamery, and the people who worked there have given us.

William C. Smith.
(Courtesy of John Cleland)

Smith Hall, what the *Lafayette Journal & Courier* once called "the best known building on the agricultural campus," underwent countless experiences, from the research conducted there that improved the worlds' nutrition for both humans and animals to the formation of lifelong friendships and long-lasting romance (as mentioned about Richard and Suzanne Belcher at the beginning of this book). Even though Smith Hall no longer serves as a dairy building and creamery, it does store the many memories of the people it helped to nourish, instruct, employ, and entertain since its construction in 1913.[38]

Sign from loading dock of creamery.
(Courtesy of John Cleland)

Appendix 1

SUPERINTENDENTS, DEPARTMENT HEADS, AND DEANS DURING THE OPERATION OF THE CREAMERY

Superintendents

M. L. Fisher — 1895 to 1901
H. E. Van Norman — 1901 to 1906
J. D. Jarvis — 1906 to 1912
H. C. Mills — 1912 to 1917
H. W. Gregory — 1917 to 1920
H. W. Fleisher — 1920 to 1921
C. O. Tuttle — 1921 to 1924
R. E. Roberts — 1924 to 1954
V. D. (Dusty) Rhodes — 1954 to 1969

Dairy Science or Animal Science Department Heads

C. S. Plumb — 1890 to 1901
H. E. Van Norman — 1901 to 1906
O. F. Hunziker — 1906 to 1917
O. J. Reed — 1917 to 1921
H. W. Gregory — 1921 to 1960

F. W. Andrews — 1961 to 1964
J. Krider — 1964 to 1969

Deans of Purdue Agriculture

John Skinner — 1907 to 1939
Harry J. Reed — 1939 to 1957
Earl Butz — 1957 to 1967
Norman J. Volk — 1967 to1968
Richard Kohls — 1968 to 1980 (creamery closed in 1969)

Appendix 2

FOUR-YEAR GRADUATES

This appendix comprises the names of people who are believed to have graduated with degrees in Dairy Manufacturing. There are no organized lists available from Purdue University in which to find the names. Various resources noted, in some cases, that the individual is a Dairy Manufacturing graduate. In other cases, only clues are left to be put together with previously known information. In some cases, assumptions were made and may be incorrect. Please remember that even the most recent of names listed are of people who graduated more than fifty years earlier in the 1960s. Apologies are offered if you are aware of an individual who graduated with this major but is not included in this list.

H. H. Madaus — Class of 1904: Even though 1904 was before Purdue started offering Dairy Manufacturing as a distinct area of specialization, students were still highly interested in studying this subject. Mr. Madaus was employed by Schlosser Bros. of Bremen following his graduation.[1] He eventually moved to Williamsport, Indiana, and became the county agent.[2]

H. C. Mills — Class of 1904: Mills went to work for Schlosser Bros. of Bremen following graduation.[3] He eventually became the Purdue Creamery superintendent.

E. C. Salisbury — Class of 1904: Salisbury, like Mills and Madaus, was employed by Schlosser Bros. of Bremen, Indiana.[4]

W. H. Klepinger — Class of 1907.[5]

W. F. Madaus — Class of 1907: Madaus was associated with Bell Springs Creamery of Abilene, Kansas.[6]

W. G. Neptune — Class of 1907.[7]

H. M. Jones — Class of 1908: After his graduation, Jones was employed by the Polk Sanitary Milk Co. of Indianapolis to supervise the company's milk plant sanitation.[8] Jones left Polk's after a brief stint and worked for another Indianapolis dairy, the Farmer's Protective Dairy Association. He later left this employment and went to work for the Purdue Dairy Department.[9]

Robert Latta — Class of 1908: Robert Latta was the son of Professor William Carol Latta. The elder Latta is recognized as the "Father of Purdue Agriculture." Following graduation, Robert Latta became the assistant butter maker for a creamery in Topeka, Indiana.[10] Latta left Topeka and went to work for the Kreechofell Dairy of Chicago, Illinois.[11] Following his stint in Chicago, Latta moved to El Paso, Texas, where he worked for the Rio Grande Valley Dairy Commission.[12]

R. T. Friend — Class of 1909: Mr. Friend entered the creamery business in Dayton, Indiana.[13] By 1932, Mr. Friend was a milk products broker in Lafayette.[14]

J. D. Groves — Class of 1910: Groves became employed as a dairy chemist for the Fairmont Creamery of Columbus, Ohio.[15]

J. L. Morehouse — Class of 1910: The Lafayette native returned to the farm.[16]

C. O. Ogle — Class of 1910: Ogle was employed as a dairy chemist by the Indiana Condensed Milk Company of Sheridan, Indiana.[17]

A. C. Rose — Class of 1911: Rose became the manager of the Evansville Pure Milk Co. ice cream plant located in Owensboro, Kentucky.[18] Following a short stay there, Mr. Rose was employed by the Fox River Butter Company of Chicago.[19]

Carl L. Wedekind — Class of 1911: Wedekind transferred into Purdue as a junior. He was a native of Louisville, Kentucky.[20]

C. R. Furnas — Class of 1912: Furnas wrote an article that was published in the April 1912 issue of the *Purdue Agriculturist*, entitled "Causes of High Prices of Butter This Winter."[21] He was employed by the Brookside Dairy near Ft. Wayne.[22]

Herman J. Kuppers — Class of 1912: Herman Kuppers was employed as a field man for the Polk Sanitary Milk Company of Indianapolis.[23]

M. P. Creath — Class of 1913: Creath was employed by the Chapin-Sacks Ice Cream Company in Washington Square, Ohio.[24]

H. R. Lochry — Class of 1913: Lochry was employed by the U.S. Department of Agriculture as a dairy manufacturing specialist covering eleven western states.[25] Mr. Lochry had one of the most unique and history-making experiences of any Purdue Dairy graduate. He helped establish and set into operation the first cheese factory in the Hawaiian Islands. In March 1926, Mr. Lochry took a leave of absence from his USDA employment to put together the equipment and operations of a cheese factory for an individual named Mr. Carter. In 1926, Mr. Carter owned one hundred head of Holstein cattle. At that time, Carter also owned what was thought to be the largest Hereford beef cattle herd in the world along with 600,000 acres of land.[26]

H. Moredock — Class of 1913: Moredock accepted a position following graduation with the Indiana Condensed Milk Company.[27] He later became superintendent of the plant located in Sheridan, Indiana.[28]

C. O. Tuttle — Class of 1913: Tuttle was employed as an efficiency man for the Schlosser Bros. Creamery of Frankfort.[29] Tuttle, like Mills, would eventually become the superintendent of the Purdue Creamery.

P. L. Borley — Class of 1914: Borley was employed by the Consumer Dairy of Jackson, Michigan.[30]

B. F. Eichinzer — Class of 1914: Eichinzer became associated with the Puritan Dairy Products Company, located on Ruckle Avenue in Indianapolis.[31]

T. T. McConnell — Class of 1914: McConnell became the manager of the Cloverleaf Dairy Company of Springfield, Missouri.[32]

G. L. Ogle — Class of 1914: Mr. Ogle became assistant manager for Ideal Pure Milk Company of Evansville.[33]

J. A. Barnett — Class of 1915: Mennen Pure Milk Co. of Lafayette employed Barnett as a milk condenser.[34]

K. G. Hassenzahl — Class of 1915: The White River Creamery of Cincinnati employed Hassenzahl as a field man.[35]

F. H. McCampbell — Class of 1915: Mr. McCampbell had offers from Evansville Pure Milk Company and Schlosser Bros of Frankfort as a butter maker at the time of his graduation. No other information has been found to determine which position he accepted.[36]

C. S. Rhode — Class of 1915: Prior to his graduation, Mr. Rhode was in negotiations with the Dairy Department at both the University of Arkansas and the University of Kansas for a creamery position.[37]

M. J. Edwards — Class of 1916: Mr. Edwards became the manager of the Blue Valley Creamery located in Columbus, Ohio.[38]

C. C. Forest — Class of 1916: Schlosser Bros. Creamery of Frankfort employed Forest as a department head.[39]

W. H. Martin — Class of 1916: Martin was employed by Creamery License Division soon after graduating. He resigned this position to work for Furnas Ice Cream Co. of Danville, Illinois.[40] In the middle of 1920; Martin turned in his resignation with Furnas and was employed by Penn State University to teach a Milk and Milk Products class.[41]

C. L. Rowe — Class of 1916: Rowe became the manager of an unnamed Chicago milk depot.[42]

B. N. Wilson — Class of 1916: Wilson was employed by the Indiana Condensed Milk Company of Indianapolis.[43]

H. F. DePew — Class of 1917: Mr. DePew was employed by the University of New Hampshire Dairy Department. He went on to be considered one of the most knowledgeable and well-known people in the ice cream industry. He worked for the General Ice Cream Corporation of Springfield, Massachusetts. DePew was a native of Auburn, Indiana.[44]

R. L. Hammond — Class of 1917: Following W. H. Martin's resignation, the Creamery License Division appointed Hammond as an assistant creamery license inspector.[45]

Irvan C. Scharf — Class of 1917: Sharf owned and managed his own milk and ice cream company in New Albany, Indiana.[46]

Harvey E. Behlmer — Class of 1918: Behlmer became the superintendent of the Blue Valley Creamery of Kansas City, Missouri.[47]

H. E. Deardorff — Class of 1918: Dearforff moved to Memphis, Tennessee, and became the manager of the Belle Meade Butter Company.[48] In 1925, Deardorff became superintendent of the Sugar Creek Creamery of Marshfield, Missouri.[49]

H. W. Fleisher — Class of 1918: In May 1924, Fleisher was manager of the Lafayette Ice Cream Company. Following his graduation in 1918, he first worked for the Creamery License Division until resigning in early 1921.[50] He then apparently went to work for Schlosser Bros. Creamery of Fort Wayne. There is speculation that he became a temporary superintendent of the Purdue University Creamery.[51] According to an article from the *Purdue Agriculturist,* dated January 1930, he was also plant manager of the Lafayette Ice Cream Company prior to becoming superintendent for Purdue. Therefore, there is a contradiction between the two articles. Other information has been found in regard to him actually being the superintendent of the Purdue Creamery. After his employment by Purdue, he worked for the Freeman Dairy Company of Flint, Michigan.[52]

J. H. Kennedy — Class of 1918: Kennedy was employed by a Sullivan, Indiana, firm, The Purity Ice Cream Co.[53]

Willard H. Martin — Class of 1918: Penn State University appointed Martin to be assistant professor of Dairy Manufacturing.[54]

V. Schwartzkoff — Class of 1918: Schwartzkoff became the general superintendent of the Blue Valley Creamery of Chicago.[55]

A. H. Henke — Class of 1919: Henke moved to Anna, Ohio, to manage a local creamery.[56]

H. J. Hickey — Class of 1919: Specializing in ice cream and butter, Hickey was joint owner of the C. L. Young Creamery of Montgomery, Alabama.[57]

E. H. Leindecker — Class of 1919: Leindecker became the assistant superintendent of the Purdue University Creamery following his graduation.[58] After leaving Purdue, Leindecker became the superintendent of the Shreveport Ice Cream Factory of Shreveport, Louisiana.[59] He later switched companies and became an ice cream maker for the Dickinson Ice Cream Company of Shreveport.[60]

E. C. Rowe — Class of 1919: Rowe was an employee of the Kokomo Sanitary Milk Company.[61]

H. A. Martin — Class of 1920: Martin became a dairy chemist for the American Milk Co. of Chicago.[62]

J. F. Beldon — Class of 1921: Beldon worked for the Alexander Creamery of Noblesville, Indiana.[63]

Burtis Elliott Horrall — Class of 1921: The Blue Valley Creamery of St. Louis, Missouri, employed Horrall as an assistant superintendent of dairy manufacturing.[64] Late in 1923, he became the manager of a Bemidji, Minnesota, plant owned by the same company.[65] While a student, the Vincennes native was also on the Dairy Products Judging Team his junior year in 1921.[66]

Neil Mckinstray — Class of 1921: While it is likely that McKinstray was a dairy manufacturing major, he and his brother Forest established and operated the McKinstray Bros. Dairy of Fishers and Indianapolis. McKinstray was a member of the Dairy Club.[67] Milk bottles from the McKinstray Dairy are rare finds for milk bottle collectors.

Ralph Preston Oyler — Class of 1921: Oyler was the alternate on the Dairy Products Judging Team in autumn of 1920. His hometown was Warsaw, Indiana.[68]

Walter Ivan Poe — Class of 1921: Poe was on the Dairy Products Judging Team his senior year. Rockville, Indiana, was his hometown.[69]

Ralph E. Roberts — Class of 1921: Roberts became one of the Purdue Dairy Science Department's most highly respected educators and dairy scientists. As a student, the Lafayette native was on the Dairy Products JudgingTteam.[70]

Frederick Karl Schmidt — Class of 1921: Schmidt was from West Lafayette, and in the autumn of 1929 during his senior year, he was on the Dairy Products Judging Team.[71]

John L. Wann — Class of 1921: Wann moved to Bloomington, Indiana, and worked as an assistant manager of the Johnson Creamery.[72]

Paul Blackburn — Class of 1922: He was another Purdue graduate hired by the Blue Valley Cream Co. to manage a local plant. Blackburn was hired for their Springfield, Illinois, location.[73] He was a Decatur, Indiana, native.[74]

George J. Brown — Class of 1922: Brown was from Milford, Indiana, and was on the Dairy Products Judging Team.[75]

Ralph Brubaker — Class of 1922: Brubaker became employed with the Furnas Ice Cream Company of South Bend.[76]

H. E. Fawcett — Class of 1922: Fawcett gained employment with the Blue Valley Creamery of New Albany, Indiana.[77]

T. B. Harrison — Class of 1922: Harrison went to the University of Tennessee as an instructor in dairy manufacturing.[78]

V. C. Manhart — Class of 1922: Manhart became a well-respected Purdue professor in Dairy Manufacturing.[79]

Ralph L. Wible — Class of 1922: Wible went to work in Owensboro, Kentucky, as an employee for the Evansville Pure Milk Company of Evansville, Indiana.[80] He was promoted to assistant manager.[81]

Hobart E. Crone — Class of 1923: Crones owned and operated a market milk plant and an ice cream plant in Sullivan, Illinois.[82]

J. T. Hoffman — Class of 1923: Mr. Hoffman accepted a position with the Schlosser Bros. Creamery located in Fort Wayne.[83]

Albert H. Jessup — Class of 1923: Jessup, of Plainfield, Indiana, joined in a partnership with his father in processing and retailing milk.[84] Jessup milk bottles are fairly easily found throughout Central Indiana.

A. N. McCutchen — Class of 1923: McCutchen was employed by the Model Ice Cream Co. of Terre Haute.[85]

L. D. Sherman — Class of 1923: Sherman represented Purdue with his teammates in St. Paul, Minnesota, in a collegiate Dairy Products Judging Contest. The Purdue team finished sixth.[86] Following graduation, he was employed by the Tip Top Creamery Company of Vincennes.[87]

John Taylor — Class of 1923: Taylor was a member of the Dairy Products Judging Team, which placed sixth at a collegiate contest in St. Paul, Minnesota on October 22, 1922.[88]

C. D. Weissert — Class of 1923: Weissert was a member of the Dairy Products Judging Team. Weissert was the second highest placing individual overall.[89] He later became manager of the Consumers Ice Cream Co. of Hammond, Indiana.[90]

H. G. Hall — Class of 1924: Sometime following graduation, Hall was employed by Creamery License Division on the Purdue campus. In late spring of 1926, Hall resigned as an assistant creamery license inspector and was then hired as a field superintendent for the Sugar Creek Creamery of Danville, Illinois.[91]

H. G. Iddings — Class of 1924: Iddings was the alternate for the Purdue Dairy Products Judging Team that placed sixth at a contest in St. Paul, Minnesota, mentioned earlier. C.O. Tuttle was the coach of the team.[92]

B. S. Mars — Class of 1924: Mars was placed in charge of the butter department for the Atlantic and Pacific Tea Company (also known as A&P) of New York City.[93]

H. C. Moore — Class of 1924: Following his graduation, Moore began working on a master's degree at the University of Minnesota.[94]

L. D. Sherman — Class of 1924: Upon graduation, Sherman accepted a position as ice cream maker for a Peru, Indiana, ice cream factory.[95]

Charles C. Walts — Class of 1924: Walts was involved in an explosion with Professor Parfitt in December 1922. No one was hurt, but windows throughout Smith Hall were blown out. He went on to do graduate work at the Connecticut Agricultural College.[96]

G. T. Badger — Class of 1925: Badger was a member of the Dairy Products Judging Team in the fall of 1922. He placed sixth overall as an individual in the national contest at the National Dairy Show.[97]

F. H. Cherry — Class of 1925: Cherry was a member of the Dairy Products Judging Team. He finished first out of thirty contestants in cheese judging in the fall of 1924. The team was the top team in cheese judging and sixth out of teams in overall dairy product judging.[98]

Clawson O. Keene — Class of 1925: Keene went to work for the Detroit Creamery following his graduation.[99] Keene was a member of the Dairy Products Judging Team in the fall of 1922. He served as the team alternate.[100]

M. J. Koldyke — Class of 1925: Koldyke was a member of the Dairy Products Judging Team.[101]

W. E. Smith — Class of 1925: Capital Dairies of Indianapolis employed Smith as a plant foreman.[102]

W. E. Stood — Class of 1925: Stood was in the market milk business in Connersville, Indiana.[103]

Rex W. Winters — Class of 1925: Mr. Winters was the manager of the Linton Milk Company of Linton, Indiana.[104]

L. F. Miller — Class of 1926: Miller became owner of Miller Ice And Dairy Company of Connersville and Cambridge City, Indiana.[105]

George Gosma — Class of 1928: All three members of the dairy cattle judging team for Purdue in the fall of 1927 were dairy product manufacturing students, Gosma included. Gosma was the second highest individual in Ayrshire judging. The team placed fifteenth out thirty-two teams.[106]

Ed Mallett — Class of 1928: Mallett, like George Gosma, was a member of the dairy cattle judging team for Purdue.[107]

Joe Scherschel — Class of 1928: Scherschel was the third member of the Purdue dairy cattle judging team, comprised entirely of dairy product students. The team placed fifteenth out of thirty-two teams.[108]

K. J. Nicholson — Class of 1929: Kenneth J. Nicholson placed third in the sophomore class Dairy Products Judging Contest.[109]

H. R. Stevens — Class of 1929: Howard Russell Stevens placed first as a judge of milk during the sophomore class Dairy Products Judging Contest.[110]

J. T. Thorne — Class of 1929: John Thomas Thorne placed second in the sophomore class Dairy Products Judging Contest. He placed second in each of the individual product categories.[111]

A. A. Beck — Class of 1930: Beck was a member of the Dairy Products Judging Team that placed second at the National Contest in the fall of 1928. He was from Thorntown, Indiana.[112]

L. J. Brand — Class of 1930: Brand was a June 1930 graduate and a Lafayette resident.[113]

N. L. Donham — Class of 1930: Donham was from Riley, Indiana, a Dairy Club member, and a June 1930 graduate.[114]

Martin Lewis Galema — Class of 1930: Galema was a January 1930 graduate. He was a member of the Dairy Club and a Lafayette native.[115] In the late 1950s, he was hired to teach the Purdue Winter Course on dairying.[116]

J. L. Harris — Class of 1930: Harris placed first overall in the sophomore Dairy Products Judging Contest for the class of 1929. He placed first in the butter category and the cheese category. Harris judged along with Thorne, Stevens, and Nicholson.[117] He later represented Purdue in the National Dairy Products Judging Contest as a member of the second-place team. Harris was from Columbia City.[118] He became field man for the Chicago Pure Milk Association.[119]

William C. Pierce — Class of 1930: Pierce was from West Lafayette and a member of the Purdue Dairy Club. He was a June 1930 graduate.[120]

Glen Pumphrey — Class of 1930: Pumphrey was a June graduate and member of the Dairy Club. He was from Connersville.[121]

Walter Virgil Roberts: Roberts's graduation date is unclear. He has been listed as graduating in both August 1930 and January 1931.[122]

Paul W. Scherschel — Class of 1930: This Bedford resident graduated in June 1930.[123]

R. C. Young — Class of 1930: Young placed second as an individual while leading the Purdue Dairy Products Judging Team to the runner-up spot at the National Dairy Products Judging Contest in the autumn of 1929. He was from West Lafayette.[124] He went on to become the assistant processing manager for the Indiana Condensed Milk Company of Lebanon, Indiana.[125]

K. H. Brown — Class of 1931: The Washington, Indiana branch of the Tip Top Creamery Co. hired Brown and made him plant manager.[126]

Alton S. Byrd — Class of 1931: Little is known of Byrd other than that he was a June 1931 graduate.[127]

Hale James Eward — Class of 1931: Eward graduated in January 1931 and was from West Lafayette.[128]

J. R. Fahl — Class of 1931: Fahl graduated in January 1931 and was from West Lafayette.[129]

John Benson Jarvis — Class of 1931: Jarvis was from Lafayette and a member of the Purdue Dairy Club.[130]

James Nolle McCarthy — Class of 1931: This graduate was a member of the Dairy Club and was a native of Paoli.[131]

Herbert Charles Snyder — Class of 1931: Snyder was from Fort Wayne.[132]

R. W. Travis — Class of 1931: Travis was employed by Swift and Co. of Keokuk, Iowa, as an ice cream maker.[133]

Joseph Merrill Trimble — Class of 1931: Trimble belonged to the Dairy Club and was from Paoli.[134]

F. N. Bennett — Class of 1932: Following graduation, Bennett was employed by Sunshine Dairy of Lafayette.[135]

Don E. Fassler — Class of 1932: Fassler returned to Kokomo and became assistant manager for Grocer's Dairy Products.[136]

C. H. Louck — Class of 1932: Louck was from Michigantown.[137]

M. D. Neuhausuer — Class of 1932: Neuhausuer returned home to Bluffton and worked in ice cream manufacturing.[138]

Wendell S. Arbuckle — Class of 1933: "Mr. Ice Cream" was one of the world's most noted ice cream researchers and authorities on ice cream. He wrote several books and articles and developed many new flavors of ice cream.

Robert Cantley — Class of 1933: Cantley was a member of the Dairy Products Judging Team.[139]

George John — Class of 1933: John was a member of the Dairy Products Judging Team for Purdue in the fall of 1932.[140]

Rueben F. Logan — Class of 1933: Logan judged on the Dairy Products Judging Team in the fall of 1932.[141]

Kenneth E. Mennen — Class of 1934: Mr. Mennen was a member of the Purdue Dairy Products Judging Team in the fall of 1932.[142] Mennen went to work for Armour and Co. Creameries of Louisville, Kentucky, following his graduation.[143]

W. M. Costello — Class of 1935: Costello was a member of the Dairy Products Judging Team for Purdue in the fall of 1934.[144] He later was the head of the Federal Milk Marketing Board for Indiana.

W. K. Fox — Class of 1935: Fox was the alternate for the Dairy Products Judging Team in the fall of 1934.[145]

John J. Million — Class of 1935: Million was a member of the Dairy Products Judging Team in the fall of 1934.[146]

W. C. Stanley — Class of 1935: Stanley was a member of the Dairy Products Judging Team in the fall of 1934.[147]

R. W. Gingrich — Class of 1937: Gingrich is of the Gingrich Dairy family of Lafayette. He was a member of the Dairy Products Judging Team in the fall of 1936.[148]

J. L. Palmer — Class of 1937: Palmer, a South Bend native, was a member of the Dairy Products Judging Team.[149]

P. H. Roll — Class of 1937: Roll was the alternate on the Dairy Products Judging Team in the fall of 1936.[150]

W. B. Schlosser — Class of 1937: Schlosser is from the famous Schlosser Bros. Creamery family. However, after graduation he chose to go to work for a creamery in Lincoln, Nebraska.[151] He was on the Dairy Products Judging Team.[152]

Floyd S. Amsler — Class of 1938: Amsler was on the 1938 Dairy Products Judging Team. He was born and raised in Rensselaer, Indiana.[153]

Karl Freese Jr. — Class of 1938: Freese, from Nappanee, Indiana, was a member of the Dairy Products Judging Team. Following graduation, he returned home to Nappanee.[154]

Thomas W. Graham — Class of 1938: Thomas Graham was on the Dairy Products Judging Team. He was a Franklin, Indiana, native.[155]

Donald W. Parsons — Class of 1938: Parsons, a Kokomo native, was a member of the Dairy Products Judging Team. Parsons started and operated the Golden Hills Dairy of Peru, Indiana.[156]

Jack Ainsworth — Class of 1939: Following graduation, Ainsworth was employed by the Carnation Milk Co. of Coshocton, Ohio.[157]

Perry Crane — Class of 1939: Crane was employed by Beatrice Meadow Gold Company of Bloomington, Indiana.[158]

Russell Knee — Class of 1939: Knee went to work for the Tip Top Creamery following his graduation.[159] The city is unknown; Tip Top had a main plant in Vincennes and another plant located in Washington, Indiana.

Edward Prusieki — Class of 1939: Following his graduation, Prusieki went to work for the White Eagle Dairy of East Chicago, Indiana.[160]

Jack Anderson — Class of 1940: The Mount Vernon, Indiana, resident was on the 1939 Dairy Products Judging Team.[161]

John Edgerton — Class of 1940: Edgerton was employed by the Borden Pure Milk and Ice Cream Company of Terre Haute following his spring 1940 graduation.[162]

Burton Page — Class of 1940: Page, from Muncie, Indiana, was on the Dairy Products Judging Team in the fall of 1939.[163]

Samuel Ramseyer — Class of 1940: Ramseyer, from West Middleton, Indiana, was a member of the Dairy Products Judging Team in the fall of 1939.[164]

Dean Wright — Class of 1940: Wright was a member of the 1939 Dairy Products Judging Team and a resident of Culver, Indiana.[165]

Gordan L. Mennen — Class of 1941: Mennen accepted a job offer from Armour and Co. of Chicago.[166]

John Henry Merhoff — Class of 1941: Merhoff graduated at the end of the summer session of 1941. The Dairy Club member and Louisville resident attended graduate school after receiving his bachelor's degree.

Walter Bauer — Class of 1942: Bauer, a Lafayette native, was a member of the winning dairy products team at the Purdue Dairy Club products judging contest in the spring of 1942.[167] He was a December 1942 graduate.[168]

Buford Butler — Class of 1942: Butler was the highest individual in the Purdue Dairy Club Dairy Products Judging Contest as a junior. Butler was from Madison, Indiana.[169] His family owned the Butler Dairy.

Max Fryback — Class of 1942: After graduating in December 1942, Mr. Fryback entered the service during World War II. Following the war, he returned home to Bluffton, Indiana, and joined his father's business, the Fryback Ice Cream Co. Mr. Fryback later served as a city councilman and mayor of Bluffton.[170]

Earl Jesse Heinton — Class of 1942: He was from Cleveland, Ohio, and a member of the Dairy Club.

Bruce C. Johnson — Class of 1942: Johnson was chosen to attend the 1942 Ice Cream Merchandising Institute in Indianapolis.[171]

V. N. Lichtenbarger — Class of 1942: This South Bend native was a Dairy Club member and graduated in December 1942.[172]

Marion C. Topliff — Class of 1942: Topliff, like classmate Bruce Johnson, attended the 1942 Ice Cream Merchandising Institute in Indianapolis.[173]

Oscar Wesson — Class of 1942: Prior to his 1942 graduation in Dairy Manufacturing, Wesson graduated from the Purdue Winter Short Course in Dairy Production. Following his 1942 graduation, he worked briefly for Kraft in Decatur, Indiana, and Bowman Dairy of Chicago as plant manager in ice cream manufacturing. Wesson then enrolled in veterinary school at Michigan State. He spent the next thirty-five years as a dairy cattle reproductive specialist in the Culver, Indiana, area.[174]

Edwin Scharf — Class of 1942: Scharf was a member of the winning junior class team for dairy products judging at the Purdue Dairy Club contest.[175] The Scharf name was well known around the Jeffersonville area for dairy manufacturing.

W. R. Calhoun — Class of 1943: Little is known of Calhoun except that he was a member of the Dairy Club and a spring 1943 graduate.[176]

J. T. Thompson — Class of 1943: Thompson, of Seymour, Indiana, graduated in August 1943. He was a member of the Purdue Dairy Club.[177] It is believed that he was a member of the Thompson Dairy family of Seymour.

Joseph Francis Shinkle — Class of 1943: This Richmond, Indiana, native was a member of the Purdue Dairy Club and graduated in December 1943.

Philip J. Rutherford — Class of 1945: Rutherford, from Springport, Indiana, went to work for Kraft following his June 1945 graduation.[178]

Willis James Harper — Class of 1946: This West Lafayette resident was a Purdue Dairy Club member and a June 1946 graduate. He went to the University of Wisconsin and earned his M.S. and Ph.D. Harper later became a professor at The Ohio State University.[179]

Herschel F. Ford — Class of 1947: Ford became head of Creamery License Division at Purdue University.[180]

J. Ralph Froedge — Class of 1947: Froedge took time out of his studies to enter the service for World War II. After graduating in January 1947, the Crawfordsville native entered graduate school.[181]

James Brandt Griffith — Class of 1947: Griffith was a January 1947 graduate.[182]

Ernie Probst — Class of 1947: After graduation, Mr. Probst managed the Waynetown Dairy before moving to Plymouth, Indiana, and working for the Plymouth Dairy. Probst passed away in 2003 at the age of 81.[183]

R. A. Allen — Class of 1948: Allen was a mid-year graduate in February 1948.[184]

Eugene T. Dragoo — Class of 1948: From Franklin, Indiana, Dragoo was a member of the Purdue Dairy Club. Following graduation, he went to work for the Kraft Cheese Co. in Palatine, Illinois.[185]

R. C. Siekman — Class of 1948: Siekman graduated in February 1948.[186]

K. V. Tost — Class of 1948: Tost graduated at mid-year in February 1948.[187] This Greenfield native was a member of the Purdue Dairy Club.[188]

William Gorman — Class of 1949: Gorman was on the Dairy Products Judging Team in the fall of 1948.[189] He became vice president and director of Kraft. Mr. Gorman was awarded the Outstanding Food Science Award presented by the Purdue University Food Science Department in 2002.[190]

J. R. Price — Class of 1949: Price was a member of the Dairy Products Judging Team and a 1949 graduate.[191]

William R. Veal — Class of 1949: Veal was a member of the Dairy Products Judging Team in the fall of 1948.[192]

Charles D. (Jack) Wesner — Class of 1949: Wesner, an Elnora native, worked for Kraft Foods for thirty-seven years.[193]

Thomas D. Selector — Class of 1950: Selector was on the Dairy Products Judging Team as a junior in the fall of 1948.[194]

Don Mather — Class of 1952: Mather, from Ridgewood, New Jersey, enrolled in graduate school at Purdue. As an undergrad, he was president of the Dairy Club.[195]

Robert Martin — Class of 1953: A Bloomington, Indiana, native, Martin had been employed by the Johnson Creamery before enrolling at Purdue. Following his graduation, he returned to the Johnson Creamery.[196]

William Jarrett — Class of 1956: Jarrett was selected as one of the top six dairy science students in the country for 1956. The Spiceland, Indiana, native was also Purdue Dairy Club president.[197]

Phil Roberts — Class of 1956: Roberts was a member of the Dairy Products Judging Team and the Dairy Club. He returned home to Indianapolis to help run the Robert's Dairy.[198]

Sam R. Winters — Class of 1956: The Linton, Indiana, native was on the Dairy Products Judging Team. He was the son of Rex Winters, owner of the Linton Dairy.[199]

Jack King — Class of 1957: King was an Indianapolis resident and Dairy Club member while he studied dairy manufacturing.[200]

Paul K. Linscott — Class of 1957: Linscott was on the Dairy Products Judging Team and was from South Bend, Indiana.[201]

Norman Napier — Class of 1957: Napier, from Fort Wayne, was on the Dairy Products Judging Team.[202]

Robert Hessong — Class of 1958: Following graduation, Bob was first employed by Hornaday Dairy in Indianapolis and then by the Borden's Company.[203]

Ross Lawrence — Class of 1958: Lawrence was a member of the Dairy Products Judging Team in the fall of 1957.[204]

Larry Stallwood — Class of 1958: Stallwood was on the Dairy Products Judging Team in 1957.[205] Following graduation, he worked in dairy management before enrolling in law school.[206]

Donald Wilkey — Class of 1958: Wilkey was a member of the Dairy Products Judging Team in the fall of 1957. He was from Kingman, Indiana.[207]

Dale L. Wolfe — Class of 1959: Wolfe was from Converse, Indiana.[208] As a dairy manufacturing student, he received his Master of Science degree in Agricultural Business from Colorado State University.[209]

Donald Barr — Class of 1960: Barr was a member of the 1959 Dairy Products Judging Team and a Logansport native.[210]

Raymond Brinegar — Class of 1960: The Bloomington resident was a member of the Dairy Products Judging Team in the fall of 1959.[211]

Billy Ray Evans — Class of 1960: Evans hailed from Indianapolis and was a member of the Dairy Products Judging Team in the fall of 1959.[212]

Donald Toloday — Class of 1960: Toloday was on the 1959 Dairy Products Judging Team, which traveled to Miami, Florida. This team was coached by R. D. Frazeur.[213]

Walter Hempenius — Class of 1962: Hempenius was a member of the Dairy Products Judging Team. The dairy manufacturing student and Lafayette native went on to receive his B. S., M. S., and Ph.D. from Purdue.[214]

Andy Azpell — Class of 1964: Azpell was one of three Dairy Manufacturing graduates in 1964. Following graduation, he was employed by Sealtest of Milwaukee.[215]

Myron Bozel — Class of 1964: Bozell accepted a job offer from Kroger after his 1964 graduation. As a student, he was a member of the Dairy Club and the Dairy Products Judging Team. In 1990, he became the head of the Creamery License Division. Bozell retired in 2007.[216]

Don Drabek — Class of 1964: Drabek was a member of the Dairy Club and the Dairy Products Judging Team.[217]

Donn L. Horein — Class of 1965.[218]

Dave Gingerich — Class of 1966: Gingerich was a Valparaiso native.[219]

Joyce Stemp — Class of 1966: Stemp was the only female that the author found to have graduated in the four-year program of Dairy Manufacturing. She was originally from Osceola, Indiana.[220]

Appendix 3

WINTER COURSE GRADUATES

The following pages consist of graduates of the Purdue Winter Couse in Creamery, Dairy Manufactures, and Dairy Manufacturing. The course titles are dependent on the era the course was offered. In the early years, the designation was Creamery, with Dairy Manufacturing next. Dairy Manufactures was used during the World War II years and then changed back to Dairy Manufacturing following the war. Most of the information listed below was taken from *The Annual Directory of Winter Course Students in Agriculture,* located in the attic of the Agriculture Administration Building. Some of the same information was found in The Virginia Kelly Karnes Archives and the Special Collections Research Center of the Purdue Library, which is located on the fourth floor of the Stewart Center. The file was marked as UA70, Box 16. Box 16 was used for the years 1906 to 1917. The folders are named with the respective year of the course. Other information was found in the *Indiana Dairy Listing Book, 3588 Dairies.* This book is a collection of known Indiana dairies since 1900, with small pieces of information about each dairy. It was compiled by this author who is a forty-five-year Indiana milk bottle collector and a member of The National Association of Milk Bottle Collectors Board of Directors.

Only the names of successful graduates are listed. Many students enrolled but did not attend. Others simply did not pass the classes or saw they would not pass and left early. Academic standards were high and quite rigorous, and it was not uncommon to have a double-digit drop-out rate. Occasionally, a student would become ill and return home.

Listed are names of students with outstanding academic accomplishments while others may be noted for historical events or from families of important Indiana dairies. Many will have dairy collectables found and valued by the public.

1906

Carl A. Anderson
Thomas A. Clark
William J. Davis
Fred Fueger
Emma E. Goldsmith
Frank Greenwell
Victor Hancock
Oscar W. Hansing
William A. Hoist
Dalia Johnston
Alice Lukens
Carl Maines
Joseph M. Martin
Christian Moeller
Herbert J. Pellam – West Lafayette

This was the first year for specific areas of study as opposed to general agriculture as had previously been taught starting in 1887. Hometowns could not be found for this class of Winter Course students except for H. J. Pellam and little is known of them. Three women were enrolled and graduated. Forty years would pass before a total of three more women would graduate. H. J. Pellam was to become butter maker for Purdue University.

1907

The seventeen names listed as students for 1907 were recorded as "dairying" with no distinction as to farm dairy production versus creamery production.

1908

Fred D. Dragoo – Muncie
Kie Kawase – Osaka, Japan
Raymond Miller – Lafayette
John Jones – Lebanon
Myron Painter – Spiceland
Cecil Schlosser – Bremen
C.L. Stahl – Ashburn, VA

Raymond Miller became qualified and licensed in milk and cream testing for an unidentified creamery in Shelbyville. Cecil Schlosser was a member of the well-known Schlosser family with creameries over much of north-central Indiana. C. L. Stahl became a dairy instructor at Virginia State University. Today Virginia State is known as Virginia Tech. He taught classes and wrote bulletins concerning the work on dairy farms and work in creameries and conducted farmer meetings across the state of Virginia.

Classes listed as being taught during the 1908 Winter Course in Dairy Manufacturing were Milk Testing, Butter Making, Butter Scoring, Creamery Accounting and Management, Creamery Machinery, Creamery Repairs, Ice Cream Manufacturing, Cheese Production, Condensed Milk, Dairy Bacteriology, and Dairy Cattle. Five individuals failed to graduate.

The opportunity for specialized instruction of both creamery and farm dairying drew students from not only Indiana but also New Jersey, Michigan, and Kentucky. There was also one Japanese student.[1]

1909

Jasper U. Aper – West Lafayette
Gorton M. Arnold – Moore Hill
Harmon C. Bender – Chrisney
Frank Connolly – Lafayette
M. E. Creane – Chicago,Illinois
Fred A. Fuhs – Huntington
John Greenham – Aurora
Cecil M. Hall – Alexandria
Walter E. Ives – Orland
A.H. Reynolds – Lewisville
J. A. Wray – New Market

Jasper Aper became a licensed milk and cream tester in West Lafayette. Jasper Aper's son Frank became well known in the dairy business as part owner of the Crawfordsville Milk Company. He also owned a creamery in Lafayette. Gorton Arnold was a milk and cream tester in Sunman, Indiana. Fred Fuhs was a licensed tester and butter maker in Otwell, Indiana. John Greenham was in the creamery business at Versailles of Ripley County. Walter Ives started the first ice cream factory in Angola. A. H. Reynolds used Lewisville as his hometown when he registered for the Winter Course, but was employed at the New Lisbon Creamery a few miles away. J. A. Wray was a member of the Wray family that operated or managed creameries, ice cream plants, and dairies in Boone, Montgomery, and Tippecanoe counties.

1910

Benjamin B. Gilmore – Moores Hill

Stephen McKay – Wirt

George H. Wells – Lafayette

Henry S. Wood – Canaan

Information is lacking about this class of four students from 1910. It is known that McKay's wife, Pearl, was enrolled to take the class, but did not attend.

1911

Prior to the 1906 class, the Winter Course had been taught for several years. There was however a lack of specialization into the different areas of agricultural practices during those years. Starting in 1906, there were Animal Husbandry, Dairy Husbandry, Creamery, and General Agriculture, each being taught individually. For the Winter Course of 1911, the faculty and administration elected to return to the non-specialized curriculum. The specialized areas of study were once again taught in 1912.

1912

Charles H. Brewer – Dyer

Claude Brown – Indianapolis

Walter D. Crothers – Avilla

Henry Fachinger – Lanesville

G. Fuentes – Santillo, Mexico

J. Kirkwood – Petroleum

C. O. Macon – Bippus

Louis E. Ness – Chicago, IL

Ora E. Newman – Petroleum

K. A. Sheets – Lowell

Albert C. Wray – Crawfordsville

What Charles Brewer did for a living is still a mystery. It is known that he was a longtime member of the Indiana State Dairy Association. Henry Fachinger managed the Lanesville Creamery. Ora Newman did creamery work in Westfield before moving to Rushville. A. C. Wray managed a milk station for Furnas Ice Cream Company in New Market before starting the Wray Brothers Creamery in Crawfordsville. He also owned the A. C. Wray Ice Cream Company of Lebanon. Henry Fachinger operated the Lanesville Creamery well into the 1940s. One-pound butter boxes can still be found from this operation.

1913

R. E. Brubeck – New Albany

Vernon B. Cain – Grabill

Ben Gruebel – Haubstadt

Heber H. Humbarger – Columbia City

H. Jessup – Camby

Cecil A. Jewett – Carmel

C. R. Sprague – Reynolds

Charles F. Woodruff – Indianapolis

R. E. Brubeck was a member of the Brubeck Dairy family of New Albany. There are embossed milk bottles from the Brubeck Dairy, but they are rare and difficult to find. Vernon Cain was a licensed cream and milk tester from Grabill of Allen County. In 1915, the Indiana Creamery License law went into effect requiring milk and cream testers to pass a state examiners test. Of nearly 400 individuals taking the test, Mr. Cain was one of only fifteen to receive a perfect score. Mr. Jewett scored 99 out of 100, while Mr. Sprague received a 98.

1914

Ray E. Adams – West Lafayette

W. A. Breyfogle – Crown Point

William Robert Glass – Madison

Glen Hoffer – Bremen

Ed L. Kiefer – Bremen

Owen Leonard – Veedersburg

I. G. Pemberton – Wolcott

Hugo O. Perner – Butler

O. A. Rice – Unknown

Albert C. Wray – Crawfordsville

Ray Adams was found to be a licensed milk and cream tester in West Lafayette for one year. No other record can be found concerning Mr. Adams and the dairy industry. Mr. Glass was the son of Frederick Glass, owner of the Frederick Glass Ice Cream Company and the Glass Dairy Company of Madison. He helped run the company after his father's death. Mr. Hoffer became a butter maker in Bremen. Ed Kiefer was also from Bremen. The Kiefer family operated a small dairy. Owen Leonard was the owner of Leonard Ice Cream Company of Veedersburg.

1915

A. U. Byers – Boonville

W. R. Gartin – Rushville

A. A. Parsell – Angola

W. O. Perigo – Center Point

B. C. Rardin – Bloomingdale

John Anthony Risch – Vincennes

C. Rosenbarger – Corydon

L. W. Shurte – Lowell

R. Truitt – Eaton, O

L. C. Vandrew – Wolcottville

Albert C. Wray – Crawfordsville

W. O. Perigo was vice-president of the Indiana Butter Makers Association. Mr. Risch was president of the Vincennes Milk Condensing Company and manufactured butter. C. A. Rosenbarger owned Rosenbarger's Restaurant in Corydon and made ice cream. Albert C. Wray completed the class for the second time as he had been a graduate of the 1914 class as well.

1916

F. H. Adams – Sandborn

W. H. Aumann – Fort Wayne

C. Davis – Carmel

O. E. Dawkins – Whittesford, TX

H. R. Harper – Martinsville

F. W. Hornfeld – Valparaiso

W. H. Leach – Sellersburg

H. M. McIntyre – New Albany

J. M. McKay – Fort Wayne

A. Nazim – Constantinople, Turkey

Ed Sebastian Risch – Vincennes

W. N. Shannon – Oakland City

M. Suleymann – Constantinople, Turkey

E. E. Thomas – Gas City

F. H. Adams was a butter maker at Peru. W. H. Leach was involved with creameries at both Sellersburg and Osgood. Ed S. Risch will be noted, not because of where he lived or for his work, but because in the winter of 1940, he sent a son to follow in his educational footsteps. Two of the students traveled from Constantinople, Turkey.

1917

H. Heady – Noblesville

Oliver T. Johnson – Chicago, IL

W. L. McClain – Orland

B. W. Mahuren – Spencer

R. C. Marine – Mooresville

D. R. Mellon – Thayer

O. E. Miller – Gilman, IL

H. C. Pond – Keokuk, IA

John P. Radford – Crawfordsville

R. P. Reagan – Coal City

R. M. Stafford – Mooresville, NC

John L. Taylor – Athol Springs, NY

E. A. Walter – Kouts

Mr. Heady was the general manager of The Quality Ice Cream Co. of Kokomo. Oliver Johnson became a Kokomo resident, eventually moving to Bloomington to take employment with the Johnson Creamery as that company's butter maker. R. C. Marine was the owner of a creamery in Tipton. Mr. Mahuren might be associated with the Mahuren Dairy of Salem, but no link between the two has been found. Mahuran dairy items are in high demand by the collecting public. Mr. Walter was employed by the Tip Top Creamery of Vincennes as butter maker. Milk bottles from the Tip Top Creamery can normally be found quite easily. John Taylor was a member of the Terre Haute Pure Milk and Ice Cream Co. Bottles from there can also be found easily in the Terre Haute area. Radford of Crawfordsville was involved for a short time with the Crawfordsville Milk Company. Some bottles from this dairy have embossed bottles with the letters of Mc & R. The *R* stands for Radford.

1918

B. C. Carter – Lafayette

E. K. Hachmeister – Homer, IL

D. T. Stalker – Bedford

The Winter Course graduates of 1918 were few in number. This was true not only for Dairy Manufacturing but also for all of the areas of agriculture taught by Purdue that winter. With the United States entering World War I, patriotism was drawing many of the young men into the armed forces who normally would have attended classes. E. K. Hachmeister was the most noted of the three graduates, as he became the butter maker for Wakarusa Co-Op Creamery. The name was later changed to Wakarusa Co-Op Dairy. Milk bottles or other dairy items from this business are exceptionally rare and difficult to find.

1919

C. H. Leauty – Indianapolis	R. J. Risch – Vincennes
E. E. Wilcken – Columbia City	G. W. Yours – Amboy

R. J. Risch was the third person with the last name of Risch to graduate from Purdue's Winter Course in Dairy Manufacturing. One might assume, at least, that they were all related by blood in some fashion. J. A. Risch attended in 1915, while E. S. Risch took the Dairy Manufacturing classes in 1916. Amboy resident G. W. Yours became fairly prominent within Indiana dairy manufacturing circles. Milk bottle caps and butter crocks with his name on them can still be found today in collectors' inventories and in the occasional antique store. The total number of graduates was still small, as World War I concluded just two months prior to the start of the 1919 classes.

The second reason this class was so small was the great influenza pandemic that spread throughout the world as soldiers returned to their native countries after the World War I peace treaty was signed. Millions of people around the world died of this viral outbreak. For a period of years, Purdue had offered one-week short courses in agriculture. However, the flu outbreak was so severe in 1919 that the university cancelled these short courses that year.

1920

G. W. Davis – Unknown	Charlie P. Myers – Marion
G. F. Green – Wakarusa	Fred Sorrels – Murpheysboro, IL
Floyd D. Mennen – Lafayette	A. E. Virgil – Bristol

Floyd Mennen was a member of the well-known Mennen family of Lafayette. Floyd was a son of Obo Mennen who started Mennen Pure Milk Company. Another son named Elmer started his own dairy called the Elmer Mennen Dairy. A milk bottle or any other item associated with these two dairies is quite difficult to find and normally very expensive. Clarence Mennen, also a son of Obo's, attended the Winter Course in 1927.

1921

R. B. Brown – Wakarusa

R. L. Davis – Ramsey

T. E. Dragoo – Muncie

R. Floyd – Broad Ripple

J. H. Kloecker – Lexington, KY

Albert F. Palmer Jr. – West New York, NJ

C. E. Pfeil – Crown Point

S. L. Risch – Vincennes

Fred Wambaugh – Elkhart

S. L. Risch was the fourth Risch to attend the Winter Course since 1915. R. L. Davis did not put his manufacturing education to use as he became a tester of cows, weighing a cow's production each day and testing it for butterfat. Mr. T. E. Dragoo returned to his dairy farm in Muncie. Also in the class of 1921 was F. W. Wambaugh. (Wambaugh did not complete the class successfully. He was a brother of Harvey Wambaugh of Elkhart. Harvey Wambaugh was a very successful owner of a dairy and a Buick automobile dealership. F. W. Wambaugh became the service manager for his brother's Elkhart Buick Sales and Service dealership. He apparently liked cars better than milk.)

1922

T. S. Braude – Silver Lake

G. R. Brooks – Pleasant Lake

V. L. Cunningham – Muncie

George I. Ellsworth – Huntingburg

C. C. Kaster – Kentucky

H. Kreamelmeyer – Missouri

G. W. Platt – Patriot

F. W. Ruppert – Angola

C. O. Story – Ashley

Lotos R. Sullivan – Salem

Among these graduates, G. I. Ellsworth becomes the best-known one, as he became owner of Ellsworth Ice Cream Company of Huntingburg. Ellsworth Dairy items are fairly easily found, particularly in southern Indiana.

1923

R. V. Atkinson – Rockport
Benjamin H. Beam – Columbus
L. C. Beck – Chicago, IL
E. F. Draut – Aurora
H. L. Hopping – Aurora
D. L. Hutchinson – Milan
J. R. Richards – Winona Lake
Wm. S. Wyss – Fort Wayne

Little is known about these graduates, although W. S. Wyss of Fort Wayne operated his own dairy after leaving a partnership with his father. The dairy was located at 3201 Lafayette Street in Fort Wayne and was called the Lafayette Dairy. Four years later in 1927, William Wyss owned his own dairy at 1141 Wildwood in Summit City. No known bottles or caps exist from either of these dairies. The Snyder Dairy had previously occupied the 1141 Wildwood address.

Estel Draut was the owner and operator of the Rising Sun Creamery Co. of Rising Sun, Indiana.

The most noted person of the class was one who did not graduate, that person being R. L. Fleischhauer of Mulberry. Mr. Fleischhauer entered class late, and left class one week before finishing. He went on to have a successful career in the dairy manufacturing industry. Today milk bottles and other items from his dairy are prized and highly priced. It was not uncommon for students not to succeed in their educational endeavors. In this class of 1923, although eight people graduated, six others besides Fleischhauer were not successful in graduating.

1924

Adelburt F. Batt – Salem
James M. Bostick – Terre Haute
Lonna E. Brown – West Lafayette
Will H. Ferguson – Salem
George W. Fisher – Orleans
Mark Galeb – Gary
John B. Grandison – Greenfield
Henry D. Iddings – Hanover
Lowell W. Johnson – Sullivan
Henry J. Kunsimiller – Cincinnati, OH
Burton F. McCormick – Lafayette
Alva F. Patterson – Fort Wayne
Francis Joseph Roberts – Lafayette
Julius E. Shad – Vevay

Two notable names from the class of 1924 were James Bostick and B. F. McCormick. The Bosticks were a dairy family from the Terre Haute and Sullivan area. Their milk bottles are on the rare side and not easily found. McCormick's milk bottles are also quite rare. In 1930, this dairy was located at 635 North Grant Street in West Lafayette. The phone number was 5741.

1925

Paul H. Gable – Mulberry

Clarence C. Hagerty – La Porte

William P. Huddleston – Unknown

Joseph S. Hughes – Irving, IL

James S. Mack – Chicago, IL

Stephen S. Martin – Orleans

Ray Roller – Fort Wayne

Hans Sailor – Independence

Arie Wiglama – Holland, Europe

Ralph Norman Van Eaton – Frankfort

Paul Gable was listed as the distinguished student of the 1925 class. This was apparently the first time this designation was used. It was not used in 1926, but started once more in 1927.

At the conclusion of this eight-week course in Dairy Manufacturing, a Dairy Products Judging Contest was held. Stephen Martin was the overall winner. He expected to make dairy manufacturing his life's work, having a year of experience in ice cream manufacturing before attending the Purdue Winter Course. Martin went to work for Johnson Creamery of Bloomington. Second place went to Ralph Van Eaton with a perfect score in the judging of butter, but he faltered slightly in other dairy products. Van Eaton returned to Frankfort to work for Schlosser Bros. Creamery.[2]

1926

D. W. Campbell – Canada

J. M. Campbell – Canada

H. M. Collins – Frankfort

Warren Hillyard – Inglefield

H. S. Jorritsma – Michigan

A. L. Morehouse – Lafayette

Michael Noonan – Fort Wayne

E. Poinsett – Fort Wayne

Delmar W. Powell – Huntingburg

P. F. Spalding – Washington

C. Stewart – Canada

The Winter Course class of 1926 saw three Canadian students as well a student from Michigan. A. L. Morehouse was from the Morehouse Milk Farm family of Lafayette. This family was one of very few enterprises in the business that called itself a "milk farm" as opposed to a dairy. Morehouse bottles and items are fairly difficult to find. Michael Noonan gained employment with the Furnas Ice Cream Co. of Fort Wayne.[3] The Furnas Ice Cream Company also had branches in Lafayette, Terre Haute, and Indianapolis.

1927

Earl L. Aders – Bristow

Leonard M. Curtis – Ohio

Floyd M. Foster – Illinois

Harold E. Grant – Jasper

Sebrig V. Hillyer – Indianapolis

Harry A. Kain – Brownsville

Brooks A. Kannapel – New Albany

Clayton G. Kessler – Crawfordsville

Clarence E. Mennen – Lafayette

Joseph B. Skomp – Lyons

Paul R. Siess – Tipton

Three states were represented in this class, as Indiana, Ohio, and Illinois all sent students. Earl Aders, Harry Kain, and Clayton Kessler were all selected as Distinguished Students. The Mennen family of Lafayette was represented by the best-known member of the Mennen family, with Clarence graduating. The Kannapel family of New Albany had a representative as well. The Kannapel Dairy would later be one of southern Indiana's largest dairies. Bottles from Elmer Mennen are nearly worth their weight in gold, while the Kannapel bottles, caps, and other dairy items are much more common and are reasonably priced.

1928

W. A. Anderson – Jeffersontown, KY

W. J. Badollet – Orleans

O. R. Carlson – Galveston, TX

Roy L. Grant – Rushville

G. S. Horton – Tipton

Paul Lottes – Tell City

W. H. Peacock – Mooresville

Theodore Tansy – Indianapolis

M. W. Teaford – Orleans

W. J. Badollet and M. W. Teaford were the outstanding students of the class of 1928, while Theodore Tansy would later become well known in the Indianapolis area as the owner of his own dairy. Tansy bottles are embossed and found fairly easily.

1929

G. F. Blake – Remington
W. F. Bogard – Bloomfield
G. R. Bruce – Lowell
H. T. Dean – Elyria, OH
Floyd B. Epperson – Hope
Charles Frederick Glass – Madison
J. E. Hill – Marion
O. Hineman – Switz City
J. F. McClellan – Scottsburg
E. R. Mills – Watertown, NJ
M. M. Moore – Crawfordsville
W. Newby – Arcola, IL
R. D. Oilar – West Lafayette
I. W. Pruss – Dillsboro
C. Ramsey – Derby
N. F. Rodgers – Greenfield
E. A. Schulz – Aurora
F. W. Tyler – Remington
L. F. Volz – Milan

J. E. Hill and R. D. Oilar were the Distinguished Students of the class. Charles Glass of Madison was the best-known dairyman on the list. He was a son of Frederick Glass, owner of the Glass Ice Cream Company and Glass Dairy. This dairy later joined with Thompson Dairy of Seymour and became Thompson — Glass Dairy. W. Newby is a person of interest, as Newby Dairy of Seymour was a well-known and highly respected dairy. No link between Mr. W. Newby and the Newby Dairy has been found.

1930

Albert A. Ackerman – Sidney, IL
J. A. Baird – Jeffersonville
C. T. Bryant – Mardon, ND
E. G. Grandison – Greenfield
P. B. Graves – Bedford
S. F. Gunter – Claypool
E. E. Hunt – Crawfordsville
C. L. Jones – Camden
Russell R. Jones – Lafayette
Fern E. Jones – Lafayette
Norman B. Kendall – Marion
Harry Lantz – Marion
L. Martinson – Bryon, OH
R. E. Reagan – Switz City
E. P. Reaves – Rushville
G. W. Russell – Mt. Vernon
H. H. Terrell – Franklin
Paul D. Thomas – Elkhart
O. B. Troyer – Kokomo

E. G. Grandison, H. H. Terrell, and O. B. Troyer were all designated as Distinguished Students. Terrill was also the winner of the Dairy Products Judging Contest for the class, while N. Kendall and R. Reagan placed second and third, respectively. Albert A. Ackerman became manager of the Borden Company shipping department located at Corpus Christi, Texas.[4] Paul Thomas of Elkhart was the owner and operator of the Thomas Dairy. Both bottles and caps are found from this dairy.

1931

Galen Shields Comer – Scottsburg

Joseph Eberly – Ligonier

Howard Elmore – Elwood

Maurice Kevin Fitzgerald – Fostoria, OH

Kenneth Franklin Hamilton – West Lafayette

Courtney Dale Heller – Marion

Lawrence L. Kays – Indianapolis

Wesley Otto Kirkpatrick – Lafayette

Emory Glen Morehouse – Lafayette

Frank G. Neikirk – Veedersburg

Glen Leon Pitser – Huntingburg

Ray Calvin Richey – Scottsburg

Bert E. Shields – Kendallville

Edward Arnold Siebenthal – Wolcott

Harry Damon Smith – Kokomo

Fred Taylor – French Lick

Ellsworth William Thomas – Fort Wayne

Delbert Board Wilson – Scottsburg

Ardra Howard Wise – Noblesville

Dudley Sylvester Wise – Noblesville

Charles W. Yates – Warren

Wesley Kirkpatrick, Bert Shields, and Dudley Wise were the outstanding students of this class. Joseph Eberly won the Diary Products Judging Contest with Courtney Heller finishing in second place. Howard Elmore had started his dairy in Elwood in 1929, two years prior to his taking classes. Round milk bottles from his dairy are highly desirable, but the square bottles are more easily found. Emory Morehouse is another member of the Morehouse Milk Farm family of Lafayette. Edward Siebenthal and his brother Emil would head up the Siebenthal Dairy of Wolcott. Their bottles are all black pyroglazed bottles with a reference to Guernsey milk imprinted on them. Their bottles do not have Wolcott on them. The milk bottle caps from Siebenthal Dairy are imprinted with Wolcott. Edward Siebenthal was the high man in the judging of butter in the Products Class Contest. Galen Comer won the ice cream division.

1932

Lawrence Herbert Brauer – Boonville

Otto Brocher – Boonville

William Harold Cook – Dana

Newell Ensly Deacon – Frankfort

Lloyd Christian Engle – West Lafayette

Wilson Thomas Handy – Greencastle

Paul Jensen – Mt. Vernon

Robert Jay King – Fort Wayne

Ezra John Klepper – Lawrenceburg

Victor Slayton Kline – Logansport

Fred White Majors – Dana

Harry Richard McGrath – Lafayette

Morgan H. Morehouse – Lafayette

Archie Garrett Murphy – West Lafayette

Edgar Frank Neuendorf – Akron

James Leroy Palmer – South Bend

Russell Floyd Phillip – Nappanee

Robert Waugh Pontzius – Columbia City

Ray Marshall Richey – Scottsburg

Joe C. Sarrazine – Fort Wayne

Floyd E. Smith – Jonesboro

Newell Deacon and Victor Kline were the Distinguished Students of the 1932 graduating Winter Course in Dairy Manufacturing. Kline was not only one of the outstanding students of the class, but also the milk bottle associated with the dairy he and his father owned and operated, Kline Dairy of Logansport, is one of the rarest bottles in Indiana. One of three known bottles recently sold for $500. Handy's Dairy of Greencastle was represented by Wilson Handy, while the Morehouse family sent another son, Morgan Morehouse. Handy's dairy bottles and other related items are in much demand around Greencastle. Floyd E. Smith worked for the Marion Pure Milk Co. Robert Pontzius was the champion dairy products judge of the class. He was high individual in ice cream judging during the contest. Newell Deacon placed second in the contest, with Fred Majors finishing third overall and first in butter judging. Paul Jensen was the high-placing individual in milk judging.

1933

Lewis Everett Beeker – West Lafayette

Harry Hanson Emig – Rose Lawn

Clive Edward Etchinson – Crawfordsville

Chester Earl Farrar – Richmond

Quinten C. Gillespie – Lafayette

R. Paul Grandison – New Castle

Earl Nelson Greider – Huntington

Harold Gunn – Veedersburg

Marvin Gustave Karch – Evansville

Fredrick Seymour Howard – Unknown

Raymond Albert Parsons – Muncie

David Foster Peyton – Hammond

Ora Everett Reinke – Wanatah

Earl Edward Renneker – Converse

Irving E. Schmeltekop – Holland

Raymond F. Schmelter – Crown Point

Clem Keller Thompson – Cortland

Howard Lee Turner – Paoli

Chester Farrar, David Peyton, and Clem Thompson were the Distinguished Students in the winter of 1933. Farrer worked for Sealtest in ice cream manufacturing for a number of years. The most noted dairymen were Ora Reinke of Wanatah and Raymond Schmelter of Crown Point. Reinke milk bottles tend to be square and can be easily found. Should you ever find a round Reinke bottle, be sure to purchase it, as they are exceptionally rare. The Schmelter bottles are difficult to find. The embossed bottles, however, are more common than the pyroglazed ones. Howard Turner, from Paoli, was of the Turner Creamery family. Milk bottles from there are exceptionally rare. Turner was also the first-place winner in the Dairy Products Judging Contest held for this class. He was also the top judge in both butter and ice cream.

1934

James Lawrence Alexander – Crawfordsville

Denzel Marvin Armstrong – Lowell

Walter David Barker – Plainfield

Earnest Earl Barkas – Columbus

Verel Henry Beemer – Sidney, OH

William Donald Blake – Remington

Richard Virgil Breininger – Defiance, OH

Lloyd Charles – LaFontaine

Earl John Carsten – Lowell

Lilburn Costello – Bluffton

T. E. Daum – Connersville

Edward Joseph Dager – Utica, OH

Joe P. Edwards – Crawfordsville

Clifton Leroy Einspahr – Lowell

Norman William Farley – Holland

Mervin Gardiner – Garden City, KS

Amos Wallace Gingerich – Fort Wayne

Robert E. Gordon – Unknown

John Everett Herron – Lafayette

Glen Hugh Klopfenstein – Woodburn

Hilbert Henry Meyer – Holland

Earl Hamilton Minor – Champaign, IL

Warren Conner Nesbitt – Oxford

Alton Keith Padgitt – Rensselaer

Vernis Allen Pate – Loogootee

Harold Vlon Pence – Swayzee

Melvin John Fred Peters – Seymour

Howard Clarence Pierce – Huron, SD

Sumner Pond – Shelbyville

Manuel William Rabourn – Acton

Norbert Revis – Lafayette

Oley Olson Robinson – Fowler

Lionel Gene Roll – Cutler

Marion William Sears – Lapel

Edgar Thomas Tanner – San Angelo, TX

Howard Franklin Waddell – Frankfort

Leo Verne Winchester – Columbus

Frank Everett Yarde – Garrett

Howard Pierce and Leo Winchester were the top students for the 1934 class. There are no known dairy owners or prominent individuals coming from this year's Winter Course. Hilbert Meyer was employed by Holland Creamery.

1935

Gail Smullen Abernathy – Rushville

Wayne Allen Buchanan – Hebron

James Leonard Camp – Lagro

John Casner Crane – Bloomfield

Allen Burnett Craven – Knightstown

Claude C. Crum – Frankfort

William Edward Davenport – Russiaville

George William Geib – Remington

Mervin Leonard Hancock – Switz City

Paul Albert Hartman – Kokomo

John Edward Harvey – South Bend

Dale Kenneth Hathaway – Garrett

Clifford Wayne Hayes – Bedford

Isaac Lawrence High – West Lafayette

George Hoerenth – Rochell, IL

Orville J. Hochmeister – Boonville

Herbert Joseph Hull – Lafayette

Frank H. Idle – Wapakoneta, OH

Stuart W. Johnson – Chicago, IL

John Frederick Kamman – Holland

Silas Dennis Kamp – Orrville, OH

Wyman Earl Kilgore – Austin, MN

Raymond Henry Kingma – Lafayette

Ralph William Kintner – Jeffersonville

Lloyd Merlin Knowlton – Texas

Alfred Lewis Kovener – Crothersville

Lowell M. Lantz – Kokomo

Robert Lee Lindsey – Rockport

Russell Acey Locke – Remington

Roy George Lucterhand – Remington

Howard Wendell McDonald – Elkhart

Woodrow Dallas McMurtrie – Veedersburg

Alvah Henry Meyer – Huntingburg

Fredrick Christian Moeller – St. Louis, MO

Norman Frank Ring – Holland

Murphey Donald Rose – Rochester

Gerald William Shively – Greenfield

Everett Shultz – Lafayette

Emil Joseph Siebenthal – Wolcott

Harold Wilfred Smith – Decatur

John Joseph Stipe – New Ross

Tenner Dale Swem – South Bend

Ralph Eugene Tyler – Remington

M. E. Weaver Jr. – Eastport, MD

Samuel W. Westerhouse – Wolcott

Oral Dwayne Wright – Worthington

David Clouser Wynn – Decatur

John Crane, Emil Siebenthal, and Ralph Tyler were all selected as Distinguished Students for the class of 1935. Notables for this class in the dairy business were Allen Craven of Knightstown and Emil Siebenthal. Mr. Craven was the son of John Craven. Together they operated the Craven's Dairy. Allen Craven was the great grandson of the original founder of Knightstown, Waitsell Munson Cary. During his lifetime, Mr. Craven held numerous offices in clubs and organizations and received many awards and civic citations for community leadership and involvement. Only small creamers remain from the Craven Dairy. These are very scarce and extremely difficult to find.

Upon graduation, Mr. Siebenthal worked for the Purdue Creamery for several years, including the delivery of dairy products on a home delivery route. He eventually matriculated back to Wolcott where he and his brother, Edward formed the Siebenthal Dairy. They milked Guernsey cows and bottled their own milk.

George Shively was involved with the family's Shively Dairy of Greenfield. Pyroglazed milk bottles are prized from this dairy. W. E. Davenport was employed by Dimmitt Dairy of Kokomo. This was one of the most modern farm dairies of the time. Bottles from there are exceptionally rare. Clifford Hayes of Bedford was an employee of the Johnson Creamery headquartered in Bloomington.

1936

George Edward Baque – Ohio

Clinton William Bastian – Ladoga

Vaughn Merton Beard – Frankfort

Bud Lucian Brown – Carthage, MO

Joe Maxwell Chezem – Frankfort

Elwood Leroy Cripe – Rossville

Harry Elmer Edison – Hamlet

Wilbur Frazell Farrall – Harvard, NE

Jay Eldon Frantz – Lafayette

Robert Eugene Fuller – Indianapolis

William Andrew Fultz – Royal Centre

Norman Henry Gocker – Elberfeld

Laurence Edward Greenburg – Newtown

Robert Herman Haan – Lafayette

John Wesley Hart – Shelbyville

Henry Leo Hasse – Beecher, IL

Robert Vincent Hatke – Lafayette

Hubert William Hooten – Paoli

John Jensen – Indianapolis

Francis Jennings Knappenberger – Mulberry

David Franklin Langeluttig – Annapolis, MD

Ralph Weldon Lowe – Frankfort

Glenn Wilbert Mays – Frankfort

James Charles Mendenhall – Marion

Frank Milicia – Cleveland, OH

Richard Ted Motter – Ney, OH

William Vertriece Myers – Van Buren

Marion Joseph Oldaker – Cambridge City

Albert Glen Peters – Frankfort

Carl Wilson Rettinger – Bourbon

Kenneth Richard Schiff – Anna, OH

Aubrey William Schoonover – Chalmers

LaMoine Arthur Schoonover – Lafayette

Theo Thayer Shaffer – Corydon

Robert Eugene Siess – Kempton

Gale Jere Stockdale – Huntington

Norman Emmett Stucker – Franklin

Dallas Maria Theobald – Andrews

Charles William Turner – Franklin

Damon Leo Vertrees – Elnora

Cecil Elwood Waggoner – Washington

Norman Richard Waldon – Franklin

John William Walker Jr. – Chesterfield, MO

Thomas Aloysius Weldon – Aurora

Thomas J. Womacks – Valparaiso

Marion Oldaker and Thomas Weldon were the outstanding students of this class for Dairy Manufacturing. Oldaker was the owner of Cambridge Ice and Dairy of Cambridge City. Robert Vincent Hatke has been traced to Lafayette and is presumed to be a relative of one of the two dairies with the Hatke name. Frank Hatke owned a dairy in Lafayette as early as 1900. Brothers to Franke Hatke, Fred and John Hatke, were known to be in business as of 1911. John Hart of Shelbyville and Floyd E. Smith of Van Buren were both involved with ice cream businesses. Dallas Theobald was the son of Chauncey Theobald, owner of Big Elm Dairy of Andrews. Little is known of the other graduates.

1937

Gerry Joseph Anthrop – Lafayette

Charles Stacey Atkins – Gary

Marlin Sylvester Bare – Greensburg

Merle Louis Baumgart – Danville, IL

Guy Geroy Bleidt – Indianapolis

Archie Robert Boggs – North Vernon

Maurice Gale Boyer – Sheridan

Richard Brooks Bullock – Chalmers

Arnold John Bussell – Remington

Leslie Harold Carlisle – Monticello

Robert Benton Christopher – Brookston

Doyle Hubert Copeland – Sharpsville

Richard L. Crow – Royal Center

William Max Dinius – Bourbon

Thomas Alva Edison – Hamlet

Russell Lowell Figert – Fort Wayne

Lyle Burdette Fordyce – West Union, IA

Lawrence Edward Greenburg – Newtown

Hubert Thomas Hendershot – Wolcott

Carl Herbert Hottenstein – Evansville

John Decator Hurless – Frankfort

Ralph Edward Johnson – Lafayette

Fern Lowell Knochel – Remington

Henry Hershall Laird – Veedersburg

Dale H. Means – Paragon

Claude Hadley McGrew – Williamsburg, IA

Ernest Frank McKim – Greensburg

John Kneale Madinger – Indianapolis

John C. Maker – Kokomo

Joseph Leslie Martin – Lafayette

George Celestine Miller – West Union, IA

William Harry Milliken – Mishawaka

Harold Maurice Moore – Eaton Rapids, MI

Edmund Stanley Nance Jr. – Brazil

John Fredrick Neuman – Cornelius, OR

Lawrence Bert Osmun – Hicksville, OH

Cloyd Jackson Pierce – Alexandria

George Albert Plummer Jr. – Peru

Kermit Franklin Sheets – Franklin

Darrel Haldean Shelmon – Remington

Earl William Spreckelson – Greensburg

Kenneth Charles Statton – Lafayette

John Delmar Tarkington – Kokomo

John Visby – Galena, IL

John Joe Walter – Lima, OH

Oscar Wesson – Richmond

James Allen Woody – Thorntown

John Beard Wright – Cayuga

Wayne Woodrow Young – Royal Center

Four students were designated as Distinguished Student for the class of 1937, the largest of the Winter Course class thus far with forty-nine graduates. The four students were Gerry Joseph Anthop, Charles Atkins Jr., John Neuman, and John Walter. Ed Nance Jr., was the most prominent name in this class, with dairy roots going back to the Nance Creamery and Nance's Dairy of Brazil. Ralph E. Johnson, of Lafayette, worked for the Purdue Creamery in 1934, 1935, 1936, and 1937. He also worked for Riverside Dairy of West Lafayette for thirteen years in addition to Rose Glen Dairy and Sunshine Dairy. These latter two dairies became prominent in the Lafayette area. During his retirement, Mr. Johnson collected milk bottles, with most of them coming from Lafayette and the surrounding area.

Prior to graduating, Gerry Anthrop worked for West Lafayette dairyman Chester Gwin, where Anthrop apparently did everything from milk cows to deliver milk to houses in a delivery truck. After graduating, he went to work for the Purdue Creamery. Mr. Anthrop worked in many areas of the creamery with jobs such as washing bottles, capping bottles, making butter, cheese, and fancy ice cream molds. Anthrop met his future wife while working at the creamery. The soon to be Mrs. Anthrop checked the retail route books of the delivery drivers for accuracy.

Another name of note from the list of graduates is that of Thomas Alva Edison. His parents must have been big fans of the inventor. He was the second person from Hamlet in two years with the last name Edison.

Charles Atkins Jr. owned and operated the Atkins Ice Cream Company. Ice cream boxes from this company are still occasionally found. Doyle Copeland was a cheesemaker at a local cheese plant.

Joseph Leslie Martin, of Lafayette, was a personal friend of the author, as both were employed as assistant creamery license inspectors. Martin was employed by the Purdue Creamery prior to its closing. The author was employed part-time by the Creamery License Division while a student in Animal Science.

1938

Leland Claire Bailey – Fort Wayne

Charles John Blum – Evansville

John Wilson Brandon – Auburn

William Dale Brown – Ohio

Leon Walter Collins – New York

Bert Russell Cook – Elkhart

James Russell Cook – Frankfort

Harry Vaughn Crites – Hartford City

Henry Paxton Darroch – Cayuga

John Elliott – Frankfort

Leslie William Gott – Spencer

Abraham Hettenvan – Brownsburg

Estel Wayne Johnson – Bloomfield

Wayne Allen Kepner – West Lafayette

Alfred Tower Krejel – Ohio

Harold Harry Larrason – Greensburg

Vernon Henry Meyer – Holland

Kermit N. Money – Shelbyville

Edwin William Rademacher – Holland

Bernard Duane Reish – Union City

Milton Lee Remling – Texas

Matthias Ringnalda – Indianapolis

Leo Eugene Risacher – Loogootee

Harold Masters Rose – Economy

Walter Miller Rose – South Carolina

James Thurston Rund – Helmsburg

Wilburn Herbert Sala – Michigan

Carl Walter Schmelter – Crown Point

Sylvanus Everett Shoemaker – Columbia City

Harold Melville Spencer – Kentucky

Donald Todd Stalker – Bedford

Calvin Raymond Steury – Decatur

Braughnman John Stone – Marion

Francis Ervin Swengel – Seymour

Nick William Tarant –Chicago

Ben Herbert Timmons – Ohio

Noble Wilson Utterback – Bargersville

M. C. Van De Walle – Mishawaka

John Howard Williams – Richmond

Charles Blum, Milton Remling, Matthias Ringnalda, and Wayne Kepner were selected as the Distinguished Students for this class. Although not having any students from Canada, as in some previous classes, the class of 1938 included students from eight different states. These states included not only Indiana and the surrounding states but also Texas, South Carolina, and New York. A total of thirty-nine students received their diplomas.

Winners of the Dairy Products Judging Contest for this class were Walter Rose, Nick Tarant, and Donald Stalker.

Four names from this list have weathered the course of history as prominent Indiana dairymen. These individuals still have milk bottles dotting collectors shelves and adorning antique store windows.

Abraham Hettenvan was the owner of Cumberland Dairies of Indianapolis and Cumberland. Cumberland is a small community just east of Indianapolis on U.S. Highway 40. Some of these milk bottles have Cumberland Dairies with Hettenvan and Sons and on them. Other bottles have Cumberland Dairies with Hettenvan and Son pyroglazed on them. This change (son without the *s*) was because one of Mr. Hettenvan's two sons was killed in combat in World War II, and after the war, all the milk bottles ordered were emblazoned with only son.

Carl Schmelter was a brother of R. F. Schmelter and worked alongside him. Both pyroglazed and embossed bottles can still be found but are difficult to locate.

Frances Swengel was a member of the family that owned Swengel Dairy of Seymour. The milk bottles from Swengel's Dairy often get overlooked because they do not have Seymour embossed on them. They are still in high demand by local residents who tend to be more familiar with them.

John Howard Williams was the owner of Williams Dairy of Richmond. His red pyroglazed milk bottles can still be easily found, particularly in eastern Indiana.

1939

James Calvin Archibald – Logansport

Richard Oliver Cramer – Winamac

Ralph William Eister – Van Wert, OH

Walter Junior Emerson Jr. – Elkhart

George English Jr. – Bicknell

Robert Donald Evans – Plymouth

Lloyd Franklin Ewing – Battle Ground

Keith Flesher – Marion

John B. Heddens – Fort Wayne

William Edward Heinl – Wapakoneta, OH

John Joseph Hrabik – Summittville

Morris Dale Isom – Bedford

Claude Harold Kissinger – La Porte

Arthur Wayne Knox – Linton

Frederick Jacob Larson – West Lafayette

Joe Allen Lewman – Bainbridge

Richard James Lewman – Bainbridge

Kenneth George Luecke – Warsaw

Donald Eugene McDowell – Logansport

Donovan Dean Mattox – Lawrenceburg, TN

Robert Howard Nesbit – Lima, OH

Orville Francis Roach – Anderson

Richard Eugene Rush – Williamsport

Jack H. Sample – Columbus

Charles Lon Schmidt – Rensselaer

Henry Lewis Snyder – Bremen

Harold W. Specht – Sheboygan Falls, WI

Mont James Starnes – Ladoga

Robert Lee Sweet – Marion

John Szabo – Genoa, OH

Lloyd Merle Tower – Evansville

Deforest Edward Voss – Harvard, IL

Robert George Wahl – Defiance, OH

Vernon Windell – West Lafayette

Orie Zook – Goshen

Thirty-five graduates received their diplomas after their eight weeks of study and work in 1939. John Szabo and Robert Wahl were chosen as the Distinguished Students for this Winter Course in Dairy Manufacturing.

Orville Roach was involved in ice cream manufacturing in Anderson, while John Hrabik returned to Summitville and was connected with the Summitville Cheese Co.

1940

Lloyd Aurther Alt – Medaryville

Maurice Lyman Busselburg – Hebron

Rex Devin Crume – Kokomo

Gerald Eugene Daniel – Tiffin, OH

Kenneth James Duncan – Lowell

Robert Floris Godby – Sheridan

William Max Gordon – Lynn

James C. Herzer – Garrett

William Kenneth Holmes – Warren

Walter Paul Hottenstein – Evansville

Richard Allen Keithley – Lowell

Francis Ervin McCracken – Montgomery

Chester Joseph Michalski – South Bend

Renos Price – Spencer

Samuel F. Porter – Pendleton

Lester Shaw Resler – Marion

Edward Sebastian Risch Jr. – Vincennes

Virgil Lloyd Robinson – Roannoke

Dale Dawson Rudicel – Huntington

Raymond Maxwell Scales – New Albany

Keith Frederick Stevens – Tipton

Wayne Tharp – Fountain City

Chester Leland Vining – Union City

James Nelson Williams – Connersville

Earl Kenneth Wilson – Urbana, IL

The class of 1940 comprised just twenty-five dairymen, a little more than half of the number in some previous years. Why this happened remains a mystery, because the other areas of study for Winter Course classes still had a very high enrollment. Gerald Daniel, Robert Floris Godby, and Chester Leland Vining made up the list for Distinguished Students for this class. Daniels was the winner of the Products Judging Contest held for the class. Mr. Vining became the owner of Purity Dairy of Union City, Indiana. Both bottles and caps can be found from Vining's dairy today.

Edward Risch Jr. followed in his father's footsteps by graduating with the Winter Course class of 1940, just as Edward Risch Sr. graduated with the class of 1916.

1941

Norman Junior Allen – Hammond

Edward Alling – Kenmore, NY

Robert William Bohlander – Elwood

Irvan Daniel Dellinger – Eaton, OH

Richard Riegle Dimitt – Kokomo

Robert Benjamin Frantz – Lafayette

Hannah Lindner – Indianapolis

Forest Royce McClain – Cincinnati, OH

Verne Willis Nelson – Ligonier

Howard Louis Robison – Shelby

Max Truman Sprunger – Berne

Robert Edward Tilley – Cincinnati, OH

Wayne Clay Younce – Marion

World War II had not yet started to draw young men away for military service, but the downward spiral continued for the class of 1941 in Dairy Manufacturing. Forest McClain earned the class's only Distinguished Student Award. He was also the first-place winner of the Dairy Products Judging Contest. Second was Verne Nelson, followed by Robert Frantz in third place.

Even though the enrollment was small, it contained some of the most popular names in the hobby of collecting Indiana dairy items, particularly milk bottles. The names Bohlander and Dimitt will get the blood flowing a little quicker through the veins of an avid collector. Bohlander bottles, particularly the round pyroglazed bottles from Elwood, command high prices with their bright orange painted features.

It seems that bottles from Dimitt's Dairy are nearly worth their weight in gold. These embossed bottles are exceptionally rare and are highly prized and sought after, not only by the residents of Kokomo, the Dimitt's hometown, but also by milk bottle collectors across the state. They did not have any pyroglazed bottles.

One extra special name of note is that of Mrs. Hannah Lindner. Mrs. Lindner is believed to have been the forth female overall and the first to take part in the Purdue Winter Course for Dairy Manufacturing since 1906 with the inception of the school's specialized creamery work. She was married to David Lindner, cofounder of Lindner Dairy and Lindner ice cream stores. She eventually became president of the two companies.

1942

Robert Adams – Fairmount

Curtis Applegate – Orleans

Robert C. Cashdollar – New Castle

Allen Gibson – Indianapolis

Melvin F. Graves – Toledo, OH

Gordon Lawrence – East Chicago, IN

Jerald Lindley – Huntington

R. C. McCoy – Osgood

Bernard Zriny – Michigan City

The 1942 Winter Course did take place, even though the United State's participation in World War II had started just a month prior. The students identified include R. C. McCoy as winner of the overall Dairy Products Judging Contest, R. C. Cashdollar, winner of the Ice Cream Judging Contest, and M. F. Graves, who was first in the Butter Judging Contest and the overall scholarship winner.[5] Enrollment for all endeavors of the Winter Course was severely diminished, with only ninety-eight students enrolled in all four areas of study, as compared to 164 students enrolled the previous year. [6]

1943

Only fifty students enrolled in the Purdue Winter Course in January 1943, with ten of these students enrolled in Dairy Production. The Dairy Manufacturing classes did not take place, either because there were no applicants or there were too few. The war continued to siphon away potential enrollees. Young men of this age were either in the military, had "essential jobs," or were physically unfit for the amount of labor involved with dairy manufacturing. Therefore, the pool of potential Winter Course students for enrollment was exceptionally small—much smaller, as is clear when you compare the 50 students to 98 in 1942, 164 in 1941, 195 in 1940, and 236 in 1939.[7] As stated earlier, the Dairy Manufacturing class was not conducted.

1944

Harry K. Paddock – Evanston, IL

John Robert Risch – Vincennes

Robert Warren Stephens – Terre Haute

The enrollment for the Dairy Manufacturing section was only three students as the total enrollment for the all the areas Purdue Winter Course studies jumped to ninety-five in January 1944.[8] Another member of the Risch family of Vincennes graduated from the Purdue Winter Course in Dairy Manufacturing.

1945

Vernon Casteel – Parkersburg, WV

Irma R. Dicke – Fort Wayne

Themis P. Morphis – Gary

Paul Robinson – Hammond

Three men and one woman graduated in Dairy Manufacturing work in 1945. Enrollment was still depressed overall as World War II was still raging. It is known that seventy-three men and four women were enrolled in the January 1945 Winter Class.[9] Irma Dicke was the fifth female to graduate in Dairy Manufacturing and only the second since 1906.

1946

Donald Ray Ayers – Crown Point

George Peters Baldwin – Evansville

James Buford Douglas – Chicago, IL

Andrew Belvin–Enochs – Bedford

Robert Ralph Gardiner – Garden City, KS

Earl William Hottenstein – Evansville

Edward John Jankowski – Michigan City

Herschell Ray Keller – Crawfordsville

Maurice William Nugent – Vincennes

Virginia Ross – North Vernon

Curt C. Sample – Hanover

James J. Sarlas – Chicago, IL

LaVerne A. Schenck – Evansville

Duane Wilson – Angola

Fourteen students were enrolled in Dairy Manufacturing. Sample was the top student in the winter class of 1946.[10] Maurice Nugent was a member of the Nugent Dairy family of Vincennes. Both milk bottles and caps can be found from this dairy. Virginia Ross was the third female to graduate since the three women from the 1906 class.

1947

Norman R. Anderson – Chicago, IL

J. Edmund Andres – San Pedros, Honduras

Don H. Burke – Decatur

Elliott Carlson – Tuscola, IL

Newell E. Carringer – La Porte

Arnold D. Chambers – Rosedale

Clarence G. Eichorst – Walkerton

William D. Evans – Chandler

Lee E. Gelarden – Tell City

William C. Gilstrap – Bedford

Irvan Gottlieb – Baltimore, MD

Roger Huter – St. Matthews, KY

Alfred Kayne – Baltimore, MD

Robert Lawson – Marion

Thomas L. McPeak – Fort Wayne

Garnet Owens – Indianapolis

James L. Palmer – South Bend

Willis Peachy – Kendallville

Carl Reisch – Union City

Ralph N. Ricklefs – Otterbein

George F. Scholl – Michigan City

John R. Seobold – Bluffton

Edward L. Smith – Huntington

Swain D. Stevenson – Carlos

Robert A. Swanson – La Porte

Arnold Tonne – Fort Wayne

Robert W. Vian – Marion

John Whybrew – Elkhart

lvin Williams – Loogootee

Paul G. Woliung – Sunman

James R. Yarling – Shelbyville

Several men in this graduating class returned to their hometowns and entered their families' noted dairy businesses, including Newell Carringer of La Porte, George Scholl of Michigan City, Robert Swanson of La Porte, Arnold Tonne of Fort Wayne, Alvin Williams of Loogootee, and James Yarling of Shelbyville.

Thomas McPeak was the Dairy Products Judging Contest winner for the Winter Course of 1947.[11]

1948

James L. Ayres – Crown Point

Edwin Bullerman – Fort Wayne

Constantine A. Carras – Hammond

John F. Catey – Peru

Robert Joe Fish – Bedford

James P. Fisher – Tell City

James E. Hendricks – Bedford

Louis W. Jenner – Rising Sun

Burleigh W. Kaster – Louisville, KY

Roger A. Klusmeyer – Valparaiso

David R. Lindner – Indianapolis

Alvin Rex Myers – Oden

Richard Eugene Sarber – Garrett

Kenneth W. Tonne – Fort Wayne

Loren Welborn – Orestes

Alethea Faith Westfall – Vincennes

Like the graduating class of 1947, the class of 1948 had several individuals from prominent dairies within the state. Edwin Bullerman, John Catey, David Lindner, Richard Sarber, and Kenneth Tonneall all went back to their hometowns and worked in their families' businesses. Alethea Westfall was the fourth female graduate.

1949

George Bartley – Glascow, KY

Carl A. Brookshire – Jonesboro

Fred A. Covalt – Muncie

Harold J. Crowe – Indianapolis

Harold L. Gribler – Van Wert, OH

Henry W. Keifer – Bremen

Lloyd B. Leamon – Stockwell

John E. Mullins – Marion

Harold A. Shelton – Crawfordsville

Harold D. Sutton – Plymouth

Theodore A. Williamson – Warren, OH

George Bartley was the highest scoring person in the Dairy Products Judging Contest held for the class, while Fred Covalt was elected president for all the Winter Course graduates.[12] Covalt was also of the Covalt Dairy family of Muncie, one of the largest and longest lasting dairies in east-central Indiana.

1950

Donald L. Bair – Oden

Donald E. Barker – Gas City

Leland E. Boyer – Elwood

James B. Burger – New Middletown

Garry G. Buttermann – Louisville, KY

George E. Carter – Owensboro, KY

Don E. Eakle – Crawfordsville

Norman H. Gocker – Elberfeld

Lewis E. Gray – Shirley

Joseph W. Hammen – Fort Wayne

Albert N. Hinchman – Connersville

Karl F. Hollenbach – Louisville, KY

Charles E. Johnson – Bloomington

James E. Kessler – La Porte

James S. Monticue – Bluffton

Charles P. Mullen – Kentland

William A. Strong – Rossville

Franklin E. Woods – Muncie

Even though enrollment for the class of 1950 rose to eighteen, little is known about any of the graduates.

1951

Elden O. Engle Jr. – Sullivan

Walter S. Ervin – St. Joe

Bruce E. Flora – Lagro

David D. Fry – West Lafayette

Robert L. King – Coshocton, OH

David Fry was of the Fry family who owned and operated Riverside Dairy on South River Road in West Lafayette.

1952

George Donovan – Elkhart

David B. Michalski – South Bend

James E. Waterhouse – Schenectady, NY

It is becoming fairly clear with only three men graduating that the end of Dairy Manufacturing Winter Courses are near.

1953

Charles W. Cox – Columbus
Waldo Crowder – Kendallville
Benjamin D'Arcy – Almont, MI
Robert W. Denny – New Albany
Frank B. Duncan – Bedford
Edward A. Hillen – New Albany
Charles R. Long – Madison
Donald E. Long – Loogootee
Granville B. Murphy – Seymour
John L. Pettit – Dupont
Garvin R. Smith – Evansville
Merritt G. Strange – Upland
Arnold D. Umbaugh – Plymouth

As enrollment dropped in Dairy Manufacturing, Purdue offered specialized two-week sessions to its students. The Market Milk session ran from January 5–16, followed by the Cottage Cheese and Buttermilk session from January 19–30. The final session, Technical Control of Dairy Products, ran from February 2–13. This arrangement allowed students to study only the areas they wanted or needed without having to be away from home or employment for a long period of time.

1954

Tilman O. Blackburn – Owensboro, KY
Gordon William Brown – Valparaiso
Otto R. Campbell – Battle Ground
Millard Ray Eldridge – Lexington, KY
Edward A. Leep – Highland
Thad Norman Murray – Marion
Glen Eugene Parker – Mooreland
Forrest Sanders Jr. – Martinsville
Garvin R. Smith – Evansville
Eugene H. Wint – Elizabethtown

Only two sessions were offered for the 1954 Winter Course. The Technical Control class was not taught in 1954.

1955

James A. Beerling – Joliet, IL

Dewitt Talmadge McClure – Indianapolis

James Manuel Rogers – Columbus

Ralph O. Stinebaugh – Lafayette

Herschel L. Whitham – Richmond

The 1955 class was the final Dairy Manufacturing Winter class to graduate.

1956

No Dairy Manufacturing classes were taught during the 1956 Winter Course program.[13]

Notes

Chapter 1. Fond Memories

1. Jack W. Wilbur, "History of Dairying," *77th Annual Report of ISDA,* 1967, 29.
2. Personal communication, December 5, 2014.
3. Personal communication, April 15, 2014.
4. Personal communication, August 12, 2014.
5. Personal communication, June 14, 2014.
6. Personal communication, July 12, 2014.
7. Personal communication, August 12, 2014.
8. Personal communication, August 12, 2014.
9. Personal communication, December 5, 2014.
10. Jack Alkire, "University Was Given Its Start by Famous Ag School," *Lafayette Journal & Courier,* September 17, 1956, 12.

Chapter 2. The Early Years, When Cream Was King

1. Purdue University Board of Trustees Meeting Minutes, 2 (September 16, 1890), 84.
2. *Purdue Agriculturist* XI, no. 2 (November 1916): 28.
3. *Purdue Alumnus* XXX, no. 7 (April 1926): 3.
4. *Bulletin of Purdue University*, 1, no. 4 (March 1901): 45.

5. Fred Whitford, *Grand Old Man of Purdue University and Indiana Agriculture* (West Lafayette: Purdue University Press, 2005), 98.
6. Jack W. Wilbur, "History of Dairying," *77th Annual Report of ISDA,* 1967, 29.
7. Purdue University Board of Trustees Meeting Minutes, 3 (June 7, 1910), 337.
8. *Purdue Alumnus* XIII, no. 7 (April 1926): 3.
9. *Purdue Debris* (1905): 69.
10. Purdue University Board of Trustees Meeting Minutes, 13 (January 20, 1943), 321.
11. *Purdue Agriculturist* VI, no. 1 (October 1911): 22.
12. *Purdue Agriculturist* XI, no. 2 (November 1916): 28.
13. *Purdue Agriculturist* XXIII, no. 4 (January 1929): 98.
14. *The Jersey Bulletin and Dairy World* XXXV, no. 41 (11 October 1916): 1592.
15. *Purdue Debris* (1905): 68.
16. *Purdue Exponent* (August 4, 1938): 5.
17. Purdue University Board of Trustees Meeting Minutes, 3 (June 6, 1899), 51.
18. *Purdue Alumnus* XIII, no. 7 (April 1926): 2.
19. *The Jersey Bulletin and Dairy World* XXXV, no. 41 (October 11, 1916): 1592.
20. *Purdue Alumnus* XIII, no. 7 (April 1926): 3.
21. *Bulletin of Purdue University I,* no. 4 (March 1901): 99.
22. *Bulletin of Purdue University V,* no. 4 (March 1905): 102–104.
23. *Purdue Agriculturist* 36, no. 2 (November 1941): 4.
24. *Purdue Agriculturist* XI, no. 2 (November 1916): 28.
25. *Bulletin of Purdue University IX,* no. 4 (March 1909): 25–26.
26. Fred Whitford. *Grand Old Man of Purdue University and Indiana Agriculture.* (West Lafayette: Purdue University Press 2005), 189.
27. *The Western World* 2, no. 2 (May 1903): 5.
28. *Purdue Alumnus* XXX, no. 7 (April 1926): 3.
29. *The Western World* 2, no. 2 (May 28, 1903): 5.
30. *Bulletin of Purdue University IV,* no. 1 (October 1903).
31. *Purdue Debris* (1905): 68.
32. Purdue University Board of Trustees Meeting Minutes, 3 (October 31, 1905), 221.

33. *Purdue Exponent* 5 (August 4, 1938).
34. *Purdue Alumnus* XIII, no. 7 (April 1926): 2.
35. O. F. Hunziker, papers from 1908–1914, https://apps.lib.purdue.edu/archon/index.php?p=collections/findingaid&id=815&q=hunziker&rootcontentid=8508#id8508.
36. *Purdue Exponent* 3 (December 1930): 1.
37. James V. Chambers, "A Century of Dairy Science at Purdue University," https://www.adsa.org/archive/centennial/posters/17684_purdue.pdf.
38. *Purdue Exponent* (December 3, 1930): 1.
39. O. F. Hunziker, papers from 1908–1914, https://apps.lib.purdue.edu/archon/index.php?p=collections/findingaid&id=815&q=hunziker&rootcontentid=8508#id8508.
40. *Purdue Alumnus* 23 (December 1935): 26.
41. *Bulletin of Purdue University* (1906): 38–39.
42. *Purdue Alumnus* XIII, no. 7 (April 1926): 2.
43. *Purdue Agriculturist* VII, no. 3 (December 1912): 52.
44. Ibid., 22.
45. Purdue Agricultural Experiment Station Circular No. 28, July 1911, Lafayette, Indiana, 18–19.
46. Purdue University Board of Trustees Meeting Minutes, 3 (October 8, 1906), 240.
47. *New York Produce Review and American Creamery* 34 (September 25, 1912): 695.
48. Purdue University Board of Trustees Meeting Minutes, 3 (June 12, 1907), 256.
49. *Purdue Agriculturist* XI, no. 2 (1916): 28.
50. *Purdue Agriculturist* XXV, no. 4 (January 1931): 79.
51. *Purdue Agriculturist* I, no. 4 (March 1907): 6.
52. *Purdue Agriculturist* III, no. 9 (June 1908): 9.
53. "The Operation of the Purdue Creamery." *Purdue Agriculturist* II, no. 5 (February 1908): 26.
54. *Purdue Agriculturist* VI, no. 1 (October 1911): 29.
55. *Purdue Agriculturist* III, no. 9 (June 1908): 9.
56. *Purdue Agriculturist* III, no. 3 (December 1908): 25–26.
57. Ibid., no. 14, 25–27.
58. http://www.elginhistory.com/eaah/eaah-ch03.htm.
59. *Purdue Agriculturist* VI, no. 1 (October 1911): 29.
60. *Purdue Agriculturist* III, no. 3 (December 1908): 25–26.

61. *Purdue Agriculturist* V, no. 8 (May 1911): 28.
62. Special pamphlet from Dairy Department.
63. *Purdue Agriculturist* 9, no. 1 (October 1914): 35–36.
64. *Purdue Agriculturist* VI, no. 9 (June 1912): 50.
65. J. H. Skinner, letter to Winthrop Stone, October 27, 1910.
66. *Purdue Agriculturist* III, no. 8 (May 1909): 23.
67. *Purdue Agriculturist* VI, no. 2 (November 1911): 29.
68. *Purdue Agriculturist* VI, no. 4 (January 1912): 39.
69. Purdue University Agricultural Experiment Station Circular No. 16, Lafayette, Indiana, July 1909, 19.
70. Ibid.
71. *Bulletin of Purdue University X*, no. 1 (October 1909): 19–21.
72. *Monthly Bulletin Indiana State Board of Health* X, no. 9 (September 1908): 106.
73. *New York Produce Review and American Creamery* 34 (September 25, 1912): 695.
74. *Purdue Debris* (1912): 38.
75. *Bulletin of Purdue University VIII*, no. 6 (February 1913): 119.
76. *Purdue Agriculturist* XVII, no. 3 (February 1923): 53.
77. *Purdue Exponent* 1 (March 5, 1926): 1.
78. *Bulletin of Purdue University VIII*, no. 6 (September 1913): 26.

Chapter 3. Smith Hall and Purdue Dairy Legends

1. *Purdue Agriculturist* VI, no. 4 (January 1912): 39.
2. *Purdue Exponent* (October 29, 1921): 4.
3. "Relatives Honor Memory of a Purdue Benefactor," *The Lafayette Journal & Courier*, October 29, 1921, 4.
4. Jack W. Wilbur, "History of Dairying," *77th Annual Report of ISDA,* 1967, 29.
5. H. W. Gregory, letter to E. L. Butz, October 14, 1958.
6. Purdue University Board of Trustees Meeting Minutes (September 14, 1912), 30.
7. Email from Bruce W. Dexter, Archive Crew Chief of Freehafer Hall, to the author.
8. www.findagrave.com/cgi-bin/fg.cgi?page=gr&GRid=25287522.
9. *The Thirty-Eighth Annual Report of Purdue University,* June 30, 1912, 56–57.

10. "Kemmer-Rabbi Residence-Mork House," www.historiccentennial.org/centennial/657_7th.htm.
11. *Purdue Agriculturist* VII, no. 3 (February 1913): 52.
12. *Purdue Agriculturist* VII, no. 4 (January 1913): 59.
13. *Purdue Agriculturist* VII, no. 5 (February 1913): 57.
14. *Fortieth Annual Report of Purdue University for the Year Ending June 30, 1914,* December 1, 1914, 6.
15. "Elite Group of Boilermakers Now Numbers 135," June 24, 2013, www.purduesports.com/genrel/062413aac.html.
16. Purdue University Board of Trustees Meeting Minutes, 4 (June 10, 1913), 42–43.
17. *Purdue Debris* (1913): 18.
18. *Purdue News* VII, no. 5 (July 1936): 12.
19. *Purdue News* VII, no. 5 (July 1936): 12.
20. J. H. Skinner, letter to E. J. Barker, December 23, 1912.
21. From original building plans found in Building Archives, Freehafer Hall.
22. *Purdue Agriculturist* VIII, no. 1 (October 1913): 57.
23. *Purdue Agriculturist* XI (1916).
24. Fred Whitford. *Grand Old Man of Purdue University and Indiana Agriculture.* (West Lafayette: Purdue University Press, 2005), 63.
25. *Purdue Alumnus* 5, no. 1 (October 1917): 26.
26. *Purdue Agriculturist* XII, no. 2 (November 1917): 152.
27. "Good Butter and Where to Get It." Purdue Faculty Publications.
28. *Purdue Agriculturist* IX, no. 1 (October 1914): 19.
29. *New York Produce Review and American Creamery* 47, no. 8 (December 25, 1918): 242.
30. *Purdue Agriculturist* IX, no. 9 (June 1915): 31.
31. *Purdue Agriculturist* XI (1916).
32. Ibid.
33. M. E. Ensminger, "Dairy Cattle Science," *The Interstate* (1971): 15.
34. "Good Butter and Where to Get It." Purdue Faculty Publications.
35. *Purdue Agriculturist* XI, no. 2 (November 1916): 28–30.
36. *Purdue Agriculturist* 3, no. 2 (November 1908): 10–11.
37. *Purdue Exponent* XXXII, no. 116 (January 28, 1921): 2.

38. O. F. Hunziker, letter to J. H. Skinner, March 20, 1916.
39. *Purdue Exponent* XXXIII, no. 75 (December 7, 1921): 1.
40. H. W. Gregory, letter to J. H. Skinner, June 17, 1926.
41. Ibid.
42. Ibid., May 21, 1926.
43. W. M. Beeson, letter to Earl Butz, February 3, 1959.
44. Ibid.
45. *Purdue Alumnus* XIII, no. 7 (April 1926): 2.
46. "Good Butter and Where to Get It." Purdue Faculty Publications.
47. *Purdue Agriculturist* XI, no. 4 (January 1917): 28–29.
48. *Purdue Agriculturist* III, no. 1 (October 1915): 22.
49. *Purdue Agriculturist* XI, no. 4 (January 1917): 28–29.
50. *Purdue Agriculturist* III, no. 1 (October 1915): 22.
51. *Purdue Agriculturist* VII, no. 3 (December 1912): 9–11.

Chapter 4. Women in the Dairy, Winter and Short Courses

1. A Course for Women in Dairying, Purdue School of Agriculture, 1918.
2. *Bulletin of Purdue University Annual Report of the President* XIX, no. 1 (December 1918): 6.
3. *Purdue Agriculturist* XII, no. 7 (April 1918): 431.
4. Charles S. Plumb, "Special Winter Course in Dairying," 1895.
5. *Bulletin of Purdue University I*, no. 4 (March 1901): 4–5.
6. Charles S. Plumb, "Special Winter Course in Dairying," 1895.
7. *Purdue Debris* (1941): 70.
8. *Purdue Agriculturist* II, no. 6 (March 1908): 11–12.
9. *Purdue Agriculturist* XX, no. 3 (December 1925): 62.
10. *Purdue Agriculturist* XVII, no. 6 (March 1923): 106.
11. Student Enrollment 1927–1928. Final Examinations 1925–1928.
12. *Purdue News* 1, no. 5 (July 1930): 10–11.
13. *Purdue Agriculturist* XIX, no. 9 (June 1925): 200.
14. H. W .Gregory, letter to H. J. Reed, November 11, 1940.
15. Ibid.
16. *Purdue Agriculturist* 42, no. 8 (May 1951): 10.
17. *Purdue Agriculturist* VIII, no.1 (October 1912): 46.

Chapter 5. The Years of H. W. Gregory and Other Dairy Greats

1. *Purdue Agriculturist* XV, no. 1 (October 1920): 38.
2. *Purdue Agriculturist* XIII, no. 1 (October 1918): 35.
3. *Purdue Agriculturist* XVI, no. 8 (May 1920): 46.
4. *The Jersey Bulletin and Dairy World* XXXIX, no. 14 (April 7, 1920): 791.
5. Purdue University Board of Trustees Meeting Minutes, 5 (April 14, 1920), 183.
6. Purdue University Board of Trustees Meeting Minutes, 5 (June 8, 1920), 200.
7. *Purdue Agriculturist* XV, no. 1 (October 1920): 38.
8. Purdue University Board of Trustees Meeting Minutes, V (October 6, 1920), 224.
9. *Purdue Agriculturist* XIX, no. 3 (December 1924): 62.
10. *Purdue Agriculturist* XVIII, no. 8 (May 1924): 164.
11. *Purdue Agriculturist* 23, no. 4 (January 1929): 98.
12. *Lafayette Journal & Courier Newspaper Index*, 1902–1952, http://www.ingenweb.org/intippecanoe/JC_f2.html.
13. *Purdue Debris* (1918): 83.
14. *Purdue University Annual Catalog*, 1920–1921, 13.
15. *Directory of Purdue Students 1921* XVIII: 5.
16. *Directory of Purdue Students 1920* XVII: 3.
17. *Directory of Purdue Students 1922* XVIV: 7.
18. Purdue University Board of Trustees Meeting Minutes, 6 (April 9, 1924), 110.
19. H. W. Gregory, letter to J. H. Skinner, September 28, 1925.
20. *Purdue Debris* (1913): 153.
21. Purdue University Board of Trustees Meeting Minutes, 5 (June 1, 1922), 489.
22. *Purdue Debris* (1922): 387.
23. *Purdue Agriculturist* XIX, no. 3 (December 1924): 62.
24. *Purdue Agriculturist* XXV, no. 4 (January 1931): 79.
25. *Purdue Debris* (1922): 387.
26. *Purdue Exponent* 4 (January 21, 1921).
27. *Purdue Agriculturist* VI, no. 1 (October 1911): 29.
28. Purdue University Board of Trustees Meeting Minutes, 5 (April 21, 1921), 464.

29. *Purdue Alumnus* 10, no. 6 (March 1923): 10.
30. *Purdue Exponent* (December 15, 1923): 1.
31. *Purdue Exponent* (December 16, 1923): 1.
32. *Purdue Agriculturist* XIX, no. 5 (February 1925): 100.
33. *Purdue Agriculturist* XIX, no. 5 (February 1925): 100.
34. *Purdue Agriculturist* XIX, no. 2 (November 1924): 31.
35. *Purdue Alumnus* XIII, no. 7 (April 1926): 4.
36. *Purdue Agriculturist* XVII, no. 6 (March 1923): 109.
37. H. W. Gregory, letter to J. H. Skinner, 31 December 1924.
38. *Purdue Agriculturist* December XX, no. 3 (1925): 68.
39. *Purdue University Bulletin-Financial Report* XXX, no. 4 (February 1930): 89.
40. *Purdue Alumnus* XIII, no. 7 (April 1926): 4.
41. *Purdue Agriculturist* XXII, no. 4 (January 1929): 98.
42. Purdue University Board of Trustees Meeting Minutes, 6 (April 14, 1926), 499.
43. *Purdue University Bulletin-Financial Report* 27, no. 9 (December 1926): 6.
44. *Purdue University Bulletin-Financial Report 28*, no. 4 (December 1927): 6.
45. *Purdue University Bulletin-Financial Report* XXIX, no. 3 (November 1928): 85.
46. *Purdue University Bulletin-Financial Report* XXX, no. 4 (February 1930): 89.
47. *Purdue University Bulletin-Financial Report* XXXI, no. 4 (December 1930): 110.
48. Purdue University Board of Trustees Meeting Minutes, 6 (April 14, 1926), 499.
49. Purdue University Board of Trustees Meeting Minutes, 6 (September 28, 1926), 586.
50. Purdue University Board of Trustees Meeting Minutes, 6 (October 13, 1926), 599.
51. Purdue University Board of Trustees Meeting Minutes, 6 (June 14, 1926), 555.
52. *Purdue Exponent* (September 28, 1928): 1.
53. http://dairyproductscontest.org/pdf/Collegiate%Dairy%20 Products%20Judging%201916-1964.pdf.
54. J. H. Skinner, letter to H. W. Gregory, August 29, 1931.
55. *Bulletin of Purdue University XXX*, no. 7 (April 1930): 119.

56. *Bulletin of Purdue University XXX*, no. 7 (April 1930): 122.
57. *Bulletin of Purdue University XXX*, no. 7 (April 1930): 170–172.
58. H. W. Gregory, letter to V. C. Freeman, July 1939.
59. H. W. Gregory, letter to O. E. Reed, August 4, 1939.
60. H. W. Gregory, letter to O. E. Reed, October 5, 1939.
61. H. W. Gregory, letter to O. E. Reed, October 2, 1939.
62. H. W. Gregory, letter to O. E. Reed, October 6, 1939.
63. H. W. Gregory, letter to O. E. Reed, October 11, 1939.
64. Ibid.
65. H. W. Gregory, letter to O. E. Reed, October 24, 1939.
66. H. W. Gregory, letter to O. E. Reed, December 28, 1939.
67. H. W. Gregory, letter to O. E. Reed, January 2, 1940.
68. H. W. Gregory, letter to O. E. Reed, February 16, 1940.
69. Purdue University Board of Trustees Meeting Minutes, 12 (January 16, 1940), 64.
70. Purdue University Board of Trustees Meeting Minutes, 12 (April 17, 1940), 146.
71. Purdue University Board of Trustees Meeting Minutes, 12 (January 16, 1940), 45–46.
72. *Purdue Exponent* 52, no. 147 (April 4, 1941): 5.
73. *Scrivener, no.* 13 (May 1940). 38

Chapter 6. The Great Depression, Forecast for the Future

1. Purdue University Board of Trustees Meeting Minutes, 8 (October 26, 1931), 282.
2. Ibid., 9 (August 1, 1933), 27–30.
3. Purdue University Board of Trustees Meeting Minutes, VIII (January 14, 1931), 74.
4. Ibid.
5. Ibid.
6. Purdue University Board of Trustees Meeting Minutes, 9 (June 11, 1934), 402.
7. Notes of interview by J. L. Albright with Earl Butz, April, 4, 2001.
8. Purdue University Board of Trustees Meeting Minutes, 24 (November 5, 1968), 841.
9. Personal conversation with Dr. Merle Cunningham, March 28, 2014.

10. *Purdue Alumnus* (May 1975): 18.
11. petemazzochi.com.
12. Rajendrani Mukhopadhyay, "The Scoop on Jerry Greenfield," *ASBMBToday,* March 2015, www.asbmb.org/asbmbtoday/201503/Features/Greenfield.
13. petemazzochi.com.
14. R. B. Stewart, letter to J. H. Skinner, January 28, 1933.
15. H. W. Gregory, letter to J. H. Skinner, April 18, 1932.
16. H. W. Gregory, letter to J. H. Skinner, January 7, 1931.
17. Ibid.
18. H. W. Gregory, letter to R. B. Stewart, February 2, 1931.
19. Purdue University Board of Trustees Meeting Minutes (October 16, 1935), 62–63.
20. Ibid.
21. H. W. Gregory, 1936–1937 Budget Request, to J. H. Skinner. n.d.
22. Purdue University Board of Trustees Meeting Minutes, 11 (June 11, 1938), 286.

Chapter 7. World War II and the Creamery

1. Purdue University Board of Trustees Meeting Minutes, 12 (January 15, 1941), 577.
2. Ibid. 13 (20 January 1943): 371.
3. *Bulletin of Purdue University Annual Report* XLIII, no. 1 (January 1943): 54.
4. *Bulletin of Purdue University Annual Report* XLII, no. 4 (April 1942): 274.
5. V. C. Manhart, letter to W. V. Lambert May 19, 1943.
6. H. W. Gregory, letter to O. E. Reed, February 9, 1943.
7. V. C. Freeman, letter to Edward C. Elliott, November 19, 1942.
8. Purdue University Board of Trustees Meeting Minutes, 13 (April 16, 1943), 422.
9. *Purdue News* 14, no. 7: 6. For Winter Course in 1944.
10. *Purdue News* 15, no. 7 (November 1944): 5. Dairy Manufactures.
11. H. T. Hamilton, letter to O. E. Reed, October 17, 1944.
12. V. C. Freeman, letter to O. E. Reed, October 30, 1944.

Chapter 8. Explosions, Accidents, Road Dust, and Sour Milk

1. https://www.adsa.org/archive/centennial/posters/17692_purduedp.pdf.
2. 1936 Budget Request from Gregory to Skinner.
3. 1940–1941 Dairy Budget.
4. *Purdue Exponent* (January 4, 1923): 1.
5. *The Nutmeg* 9 (1925): 26.
6. Accident Reports from Roberts to Gregory, 1941.
7. Purdue University Board of Trustees Meeting Minutes, 5 (June 12, 1923), 684.
8. Purdue University Board of Trustees Meeting Minutes, 6 (July 29, 1924), 192.
9. H. W. Gregory, letter to J. H. Skinner, December 17, 1925.
10. Ibid, April 11, 1928.
11. J. H. Skinner, letter to H. W. Gregory, May 1, 1928.
12. Ibid., May 28, 1928.
13. Ibid., July 8, 1927.
14. Purdue University Board of Trustees Meeting Minutes, 11 (January 18, 1939), 577–578
15. Purdue University Board of Trustees Meeting Minutes, 11 (18 January 1939), 577.

Chapter 9. Employees of the Creamery

1. *Purdue Debris* (1905): 69.
2. *Purdue University 19th Annual Report of the Agricultural Experiment Station for Year Ending June 30, 1906,* 20.
3. *Purdue Debris* (1912): 59.
4. John D. Jarvis, ed., *Twenty-First Annual Report of the Indiana State Dairy Association,* 10.
5. *Purdue Agriculturist* II, no. 1 (October 1907): 20.
6. Frederick Whitford. *For the Good of the Farmer*, (West Lafayette, IN: Purdue University Press, 2013), 214.
7. 1929–1930 Creamery Budget Request. From Gregory to Skinner, June 1929.
8. 1930–1931 Creamery Budget Request. From Gregory to Skinner, June 1930.
9. H. W. Gregory, letter to J. H. Skinner, December 17, 1931.
10. 1937–1938 Creamery Budget Request. From Gregory to Skinner, June 1937.

11. H. W. Gregory, letter to J. H. Skinner, n.d.
12. 1946–1947 Budget Request. From Gregory to Skinner.
13. H. W. Gregory, letter to J. H. Skinner, n.d.
14. Purdue University Staff Directory 1956–1957, 41.
15. Purdue University Board of Trustees Meeting Minutes, 10 (October 19, 1937), 808.
16. Purdue University Board of Trustees Meeting Minutes, 13 (October 22, 1943), 558.
17. Purdue University Board of Trustees Meeting Minutes, 13 (October 18, 1944), 833.
18. H. W. Gregory, letter to J. H. Skinner, June 1941.
19. Ibid., June 1944.
20. Bob Kreibel, "Reminiscing about Dairies," *Lafayette Journal & Courier,* May, 27, 2001, C3.
21. Purdue University Board of Trustees Meeting Minutes, 14 (January 21, 1948), 753.
22. Ernie Probst, unpublished manuscript.
23. *Lafayette Journal & Courier,* May 27, 2001, C3.
24. Personal friend of author.
25. Purdue University Board of Trustees Meeting Minutes, 13 (October 21, 1942), 222.
26. Purdue Board of Trustees Meeting Minutes, 13 (October 22, 1943), 533.
27. http//www.legacy.com/obituary/jconline/obituary.aspx?pid=173261969.
28. http://www.genlookups.com/in/webbbs_config.pl/noframes/read/650.

Chapter 10. The Years of Rhodes, Babel, and Parmalee

1. School of Ag. Announcements for 1969–1970, 123.
2. Purdue University Board of Trustees Meeting Minutes, 13 (April 20, 1944), 698.
3. Purdue University Board of Trustees Meeting Minutes, 6 (November 16, 1923), 48.
4. Purdue University Board of Trustees Meeting Minutes, 13 (May 10, 1945), 909.
5. Personal communication, August 12, 2014.
6. Purdue University Board of Trustees Meeting Minutes, 13 (May 10, 1945), 909.

7. Purdue University Board of Trustees Meeting Minutes, 14 (March 12, 1947), 480.
8. Purdue University Board of Trustees Meeting Minutes, 14 (October 24, 1947), 725.
9. Purdue University Board of Trustees Meeting Minutes, 14 (October 20, 1948), 913.
10. *Purdue Agriculturist* 41, no. 5 (February 1950): 12.
11. Personal communication, December 16, 2015.
12. Purdue University Board of Trustees Meeting Minutes, 15 (April 21, 1948), 800.
13. Ibid.
14. Purdue University Board of Trustees Meeting Minutes, 15 (April 19, 1950), 397.
15. Purdue University Board of Trustees Meeting Minutes, 15 (January 17, 1951), 664.
16. Purdue University Board of Trustees Meeting Minutes, 15 (December 19, 1951), 921.
17. Purdue University Board of Trustees Meeting Minutes, 15 (October 17, 1951), 831.
18. Purdue University Board of Trustees Meeting Minutes, 16 (January 20, 1954), 407.
19. Purdue University Board of Trustees Meeting Minutes, 22 (May 5, 1965), 400.
20. 1953–1954 Dairy Department Budget Requests, March 25, 1953. H. W. Gregory to Dean H. J. Reed.
21. Purdue University Board of Trustees Meeting Minutes, 16 (January 20, 1954), 407.
22. Personal communication, April 14, 2014.
23. Personal communication, June 14, 2014.
24. Notes on conversation with Lorraine Albright, April 12, 2014; also recollections of author.
25. James V. Chambers, "A Century of Dairy Science at Purdue University" (2006), American Dairy Science website, www.adsa.org/archive/centennial/posters/17684 (accessed March 14, 2014).
26. Patent website, www.patententimages.storage.googleapis.com/pdf/us2971847.pdf (accessed April 19, 2014).
27. *Purdue Agriculturist* 48, no. 1 (November 10, 1956): 12.

28. Obituary for Fred Babel, *Journal of Dairy Science* 81 (1998): 1801–1802.
29. Purdue University Board of Trustees Meeting Minutes, 22 (October 20, 1964), 161.
30. Ibid.
31. Purdue University Board of Trustees Meeting Minutes, 23 (March 1, 1967), 583.
32. James V. Chambers, "A Century of Dairy Science at Purdue University" (2006), American Dairy Science website, www.adsa.org/archive/centennial/posters/17684 (accessed March 14, 2014).
33. Ibid.
34. *Purdue Alumnus* 65, no. 9 (summer 1978): 8.
35. Personal conversation, September 25, 2015.
36. Personal conversation, January 7, 2015.
37. *The Purdue Agriculturist* 48, no. 1 (November 10, 1956): 12.
38. Original recipe located in Agricultural Alumni Office, Pfendler Hall, Purdue University.
39. Purdue Swiss-Type Cheese, Purdue University Experiment Station, Lafayette, Indiana.
40. *The Purdue Agriculturist* 39, no. 5 (March 1948): 6.
41. *The Purdue Agriculturist* 59, no. 2 (November 1967): 2.
42. *Purdue Alumnus* 57, no. 2 (November 1969): 22.
43. *Purdue Alumnus* 58, no. 1 (September 1970): 28.
44. *Purdue Alumnus* 59, no. 1 (September 1971): 31.
45. *Purdue Alumnus* 60, no. 1 (September 1972): 26.
46. Personal communication, December 18, 2015.

Chapter 11. A New Creamery?

1. H. W. Gregory, letter to E. L. Butz, N. J. Volk, and V. C. Freeman, October 14, 1958.
2. Ibid., April 1, 1959.
3. E. L. Butz, letter to H. W. Gregory, June 3, 1959.
4. Purdue University Board of Trustees Meeting Minutes, 19 (June 4, 1960), 694.
5. F. C. Andrews, letter to administrators, November 23, 1960.
6. H. W. Gregory, letter to E. L. Butz, V. C. Freeman, and N. J Volk, December 23, 1960.

7. F. N. Andrews, letter to V. D. Rhodes, F. J. Babel, and B. J. Liska, September 22, 1961.
8. Purdue University Board of Trustees Meeting Minutes, 20 (July 25, 1962), 806.

Chapter 12. Casualties Occur

1. "Purdue's Research Creamery Doesn't Make the Butter It Sells," *The Indianapolis Star*, 28 September 1962, 1.
2. Purdue University Board of Trustees Meeting Minutes, 23 (March 1, 1967), 583.
3. E. L. Butz, letter to V. C. Freeman, October 10, 1960.
4. V. C. Freeman, folder sent to E. L. Butz, January 21, 1958.
5. F. D. Odell, paper sent to V. C. Freeman, n.d.
6. F. D. Odell, letter to V. C. Freeman, July 15, 1960.
7. June 1963 purchase orders, June 1, 1963.
8. Purdue University Board of Trustees Meeting Minutes, 21 (October 23, 1963), 549.
9. Purdue University Board of Trustees Meeting Minutes, 22 (October 20, 1964), 191.
10. Ibid.
11. Purdue University Board of Trustees Meeting Minutes, 21 (May 31, 1963), 262.
12. L. J. Freehafer, letter to School of Agriculture and Animal Science Administrations, July 19, 1963.
13. V. C. Freeman, letter to J. L. Krider and L. J. Freehafer, July 19, 1963.
14. Notes of meeting by V. C. Freeman, of Purdue Agriculture Administration. "Record of Discussion and Recommendations Concerning Termination of Purdue Creamery Retail Routes," July 17, 1963.
15. "Remembering Jobs on a Dairy," *The Lafayette Journal & Courier,* May 20, 2001, C3.
16. V. C. Freeman, letter to J. L. Krider and L. J. Freehafer, July 19, 1963.
17. F. D. Odell, letter to twelve campus administrators, May 14, 1964.
18. Personal communication, January 20, 2015.
19. *Purdue Agriculturist* 56, no. 4 (March 1965): 7.

20. E. L. Butz, letter to Tom Maynard, August 7, 1961.
21. *Purdue Agriculturist* 58, no. 5 (April 1967): 4.
22. Purdue University Board of Trustees Meeting Minutes, 25 (April 24, 1969), 152.
23. Purdue University Board of Trustees Meeting Minutes, 24 (January 16, 1968), 204–205.
24. Personal communication, 11 May 2015.
25. Purdue University Board of Trustees Meeting Minutes, 25 (April 1969), 152.
26. *Purdue Alumnus* 56, no. 9 (summer 1969): 42.
27. Personal communication, August 12, 2014.
28. Personal communication, December 16, 2015.
29. Purdue University Board of Trustees Meeting Minutes, 25 (June 6, 1969), 319.
30. Purdue University Board of Trustees Meeting Minutes, 22 (October 20, 1964), 191.
31. Notes by J. L. Albright of interview with Earl Butz, April 4, 2001.
32. *Purdue Alumnus* 65, no. 7 (April 1978): 4.
33. Purdue University Board of Trustees Meeting Minutes, 27 (November 8, 1972), 8.
34. *Purdue Alumnus* 65 no. 7 (April 1978): 4.
35. *Lafayette Journal & Courier*, September 7, 1991, n.v., n.p.
36. *Lafayette Journal & Courier*, March 21, 1992, n.v., n.p.
37. "Relatives Honor Memory of a Purdue Benefactor," *Lafayette Journal & Courier*, October 29, 1921, 4.
38. *Lafayette Journal & Courier*, September 18, 1956, 12.

Appendix 2. Four-Year Graduates

1. *Purdue Agriculturist* I, no. 7 (June 1907): 14.
2. *Purdue Agriculturist* XVIII, no. 4 (January 1924): 74.
3. *Purdue Agriculturist* I, no. 7 (June 1907): 14.
4. Ibid.
5. *Purdue Agriculturist* I, no. 2 (January 1907): 5.
6. *Purdue Agriculturist* XX, no. 4 (January 1926): 100.
7. *Purdue Agriculturist* I no. 2 (January 1907): 5.
8. *Purdue Agriculturist* III, no. 3 (December 1908): 24.
9. *Purdue Agriculturist* 3, no. 7 (April 1909): 24.

10. *Purdue Agriculturist* III, no. 3 (December 1908): 24.
11. Ibid.
12. *Purdue Agriculturist* XX, no. 8 (April 1926): 170.
13. *Purdue Agriculturist* XX, no. 3 (December 1925): 62.
14. *Purdue Agriculturist* 27, no. 1 (October 1932): 6.
15. *Purdue Agriculturist* XX, no. 2 (December 1925): 45.
16. *Purdue Agriculturist* XX, no. 8 (April 1926): 1971
17. *Purdue Agriculturist* VII, no. 1 (October 1912): 57.
18. *Purdue Agriculturist* XVIII, no. 3 (December 1923): 36.
19. *Purdue Agriculturist* VIII, no. 5 (February 1974): 48.
20. College of Agriculture Alumni information.
21. *Purdue Agriculturist* VI, no. 7 (April 1912): 46.
22. *Purdue Agriculturist* VII, no. 1 (October 1912): 57.
23. *Purdue Agriculturist* XI, no. 6 (January 1917): 70.
24. *Purdue Agriculturist* XX, no. 3 (December 1925): 62.
25. *Purdue Agriculturist* XVII, no. 3 (February 1923): 53.
26. *Purdue Exponent* (March 5, 1926): 1.
27. *Purdue Agriculturist* XI, no. 2 (November 1916): 28
28. *Purdue Agriculturist* XVII, no. 5 (February 1923): 87
29. *Purdue Agriculturist* XIV, no. 1 (October 1919): 38.
30. *Purdue Agriculturist* XIX, no. 9 (June 1925): 200.
31. *Purdue Agriculturist* XX, no. 3 (December 1925): 62.
32. *Purdue Agriculturist* XVI, no. 7 (April 1922): 115.
33. *Purdue Agriculturist* XXVI, no. 3 (December 1931): 37.
34. *Purdue Agriculturist* XIX, no. 8 (May 1925): 176.
35. *Purdue Agriculturist* X, no. 6 (March 1916): 44.
36. O. F. Hunziker, letter to J. H. Skinner, June 8, 1915.
37. Ibid.
38. *Purdue Agriculturist* XVII, no. 7 (April 1927): 127.
39. *Purdue Agriculturist* XVI, no. 9 (June 1922): 150.
40. *Purdue Agriculturist* XIV, no. 4 (January 1920): 212.
41. *Purdue Agriculturist* XV, no. 1 (October 1920): 18.
42. *Purdue Agriculturist* XIII, no. 1 (October 1923): 19.
43. *Purdue Agriculturist* XIX, no. 9 (June 1925): 202.
44. *Purdue Agriculturist* LXXIII, no. 8 (May 1951): 10.
45. *Purdue Agriculturist* XIV, no. 4 (January 1920): 212.
46. *Purdue Agriculturist* XVIII, no. 9 (June 1924): 181.
47. *Purdue Agriculturist* XVIII, no. 2 (November 1923): 32.
48. *Purdue Agriculturist* XVIII, no. 9 (June 1924): 181.

49. *Purdue Agriculturist* XX, no. 3 (December 1925): 62.
50. Purdue University Agricultural Experiment Station: Circular No. 81, 1918; Circular No. 93, 1919, 2; Circular No. 97, 1920, 2; Circular No. 103, 1921, 2.
51. *Purdue Agriculturist* XVIII, no. 8 (May 1924): 164.
52. *Purdue Agriculturist* XXIV, no. 4 (January 1930): 88.
53. *Purdue Agriculturist* XIX, no. 9 (June 1925): 202.
54. *Purdue Agriculturist* XVIII, no. 2 (November 1923): 32.
55. *Purdue Agriculturist* XVIII, no. 9 (June 1924): 181.
56. *Purdue Agriculturist* XVIII, no. 5 (February 1924): 103.
57. *Purdue Agriculturist* XVII, no. 9 (June 1923): 170.
58. *Purdue Agriculturist* XIV, no. 1 (October 1919): 38.
59. *Purdue Agriculturist* XVIII, no. 2 (November 1923): 32.
60. *Purdue Agriculturist* XIX, no. 5 (February 1925): 100.
61. *Purdue Agriculturist* XX, no. 9 (May 1926): 193.
62. *Purdue Agriculturist* XIX, no. 9 (June 1925): 202.
63. *Purdue Agriculturist* XX, no. 4 (January 1926): 100.
64. *Purdue Agriculturist* XVII, no. 2 (November 1922): 32.
65. *Purdue Agriculturist* XVIII, no. 4 (January 1924): 75.
66. *Purdue Debris* (1921): 99.
67. *Purdue Debris* (1921): 107.
68. *Purdue Debris* (1921): 113.
69. *Purdue Debris* (1921): 115.
70. *Purdue Debris* (1921): 119.
71. *Purdue Debris* (1921): 121.
72. *Purdue Agriculturist* XVIII, no. 5 (February 1924): 103.
73. *Purdue Agriculturist* XVIII, no. 2 (November 1923): 32.
74. *Purdue Debris* (1922): 70.
75. *Purdue Debris* (1922): 72.
76. *Purdue Agriculturist* XX, no. 3 (December 1925): 62.
77. Ibid.
78. *Purdue Agriculturist* XVII, no. 3 (December 1922): 53.
79. *Purdue Agriculturist* XVIII, no. 6 (March 1924): 124.
80. *Purdue Agriculturist* XVII, no. 6 (March 1923): 106.
81. *Purdue Agriculturist* XVIII, no. 5 (February 1924): 10.
82. *Purdue Agriculturist* XIX, no. 8 (May, 1925): 176.
83. *Purdue Agriculturist* XX, no. 3 (December1925): 62.
84. *Purdue Agriculturist* XX, no. 3 (December1925): 62.
85. *Purdue Agriculturist* XIX, no. 9 (June 1925): 202.

86. *Purdue Agriculturist* XVII, no. 2 (November 1922): 39.
87. *Purdue Agriculturist* XX, no. 2 (November 1925): 45.
88. *Purdue Agriculturist* XVII, no. 2 (November 1922): 39.
89. *Purdue Agriculturist* XVII, no. 2 (November 1922): 39.
90. *Purdue Agriculturist* XIX, no. 8 (May 1925): 176.
91. *Purdue Agriculturist* XX, no. 9 (June 1926): 217.
92. *Purdue Agriculturist* XVII, no. 2 (November 1922): 39.
93. *Purdue Agriculturist* XXVI, no. 3 (December 1931): 37.
94. *Purdue Agriculturist* XVIII, no. 9 (June 1924): 181.
95. Ibid.
96. *The Nutmeg* 9 (1925): 26.
97. *Purdue Agriculturist* XVII, no. 2 (November 1922): 39.
98. Ibid.
99. *Purdue Agriculturist* XXI, no. 5 (February 1927): 110.
100. *Purdue Agriculturist* XVII, no. 2 (November 1922): 39.
101. Ibid.
102. *Purdue Agriculturist* 27, no. 1 (October 1932): 6.
103. *Purdue Agriculturist* XX, no. 9 (May 1926): 204.
104. *Purdue Agriculturist* XXIV, no. 3 (December1929): 62.
105. *Purdue Agriculturist* XXVI, no. 2 (November 1931): 21.
106. *Purdue Agriculturist* XXII, no. 2 (November 1927): 42.
107. Ibid.
108. Ibid.
109. *Purdue Agriculturist* XXI, no. 6 (March 1927): 139.
110. Ibid.
111. Ibid.
112. *Purdue Agriculturist* XXIII, no. 4 (January 1929): 88.
113. Purdue Archives, Student List 1938–1939.
114. Ibid.
115. Ibid.
116. Purdue University Board of Trustees Meeting Minutes, XVII (April 17, 1957), 735.
117. *Purdue Agriculturist* XXI, no. 6 (March 1927): 139.
118. *Purdue Agriculturist* XXIII, no. 4 (January 1929): 88.
119. *Purdue Agriculturist* 26, no. 7 (April 1932): 101.
120. Purdue Archives, Student List 1938–1939.
121. Ibid.
122. Ibid.
123. Ibid.

124. *Purdue Agriculturist* XXIII, no. 4 (January 1929): 88.
125. *Purdue Agriculturist* XXVI, no. 2 (November 1931): 21.
126. *Purdue Agriculturist* 26, no. 4 (January 1932): 55.
127. Purdue Archives, Student List 1938–1939.
128. Ibid.
129. Ibid.
130. Ibid.
131. Ibid.
132. Ibid.
133. *Purdue Agriculturist* XXVI, no. 8 (May 1932): 117.
134. Purdue Archives, Student List 1938–1939.
135. *Purdue Agriculturist* 27, no. 1 (October 1932): 6.
136. *Purdue Agriculturist* XXVIII, no. 6 (March 1934): 51.
137. Purdue Archives, Student List 1938–1939.
138. Ibid.
139. *Purdue Agriculturist* 27, no. 1 (October 1932): 5.
140. Ibid.
141. Ibid.
142. *Purdue Exponent* (November 12, 1932): 1.
143. *Purdue Agriculturist* 29, no. 4 (January 1935): 39.
144. *Purdue Agriculturist* 29, no. 3 (December 1934): 27.
145. Ibid.
146. Ibid.
147. Ibid.
148. *Purdue Agriculturist* 31, no. 1 (October 1936): 99.
149. Ibid.
150. *Purdue Agriculturist* 31, no. 1 (October 1936): 20.
151. *Purdue Agriculturist* 31, no. 6 (March 1937): 97.
152. *Purdue Agriculturist* 31, no. 1 (October 1936): 99.
153. "For the Reading," *Times of Hammond,* October 21, 1937, 77.
154. Ibid.
155. Ibid.
156. Ibid.
157. *Purdue Agriculturist* 34, no. 1 (October 1939): 11.
158. *Purdue Agriculturist* 34, no. 3 (December 1939): 2.
159. *Purdue Agriculturist* 34, no. 4 (January 1940): 13
160. *Purdue Agriculturist* 34, no. 7 (February 1940): 8.
161. *Purdue Agriculturist* 34, no. 1 (October 1939): 16.
162. *Purdue Agriculturist* 34, no. 4 (January 1940): 13.

163. *Purdue Agriculturist* 39, no. 1 (October 1939): 16.
164. Ibid.
165. Ibid.
166. *Purdue Agriculturist* 35, no. 9 (June 1941): 15.
167. *Purdue Agriculturist* 36, no. 8 (May 1942): 7.
168. Purdue Archives, Student List 1938–1939.
169. *Purdue Agriculturist* 36, no. 8 (May, 1942): 6.
170. Purdue Archives, Student List 1938–1939.
171. *Purdue Agriculturist* 36, no. 8 (May 1942): 7.
172. Purdue Archives, Student List 1938–1939.
173. *Purdue Agriculturist* 36, no. 8 (May 1942): 7.
174. *History of Marshall County*, Marshall County Historical Society, (1986) 452–453.
175. *Purdue Agriculturist* 36, no. 8 (May 1942): 7.
176. Purdue Archives, Student List 1938–1939.
177. Ibid.
178. Ibid.
179. Ibid.
180. H. W. Gregory, letter to Director H. J. Reed, April 4, 1949.
181. *Purdue Agriculturist* XXXVIII, no. 5 (February 1947): 9.
182. Purdue Archives, Student List 1938–1939.
183. Author's recollections. Personal friend of author.
184. *Purdue Agriculturist* XXXIX, no. 5 (March 1948): 6.
185. Purdue Archives, Student List 1938–1939.
186. *Purdue Agriculturist* 39, no. 5 (March 1948): 6.
187. Ibid.
188. *Purdue Debris* (1948): 115.
189. *Purdue Agriculturist* 40, no. 5 (February 1949): 17.
190. "Department of Food Science Alumni and Friends Recognitions," https://ag.purdue.edu/foodsci/Pages/Alumni-and-friends-Recognition.aspx.
191. *The Purdue Agriculturist* 40, no. 5 (February 1949): 17.
192. Ibid.
193. http://obituaries.washtimesherald.com/stories/obituaries/charles-jack-wesner-19.
194. *Purdue Agriculturist* 40, no. 5 (February 1949): 17.
195. *Purdue Agriculturist* 43, no. 8 (May 1952): 12.
196. *Purdue Agriculturist* 44, no. 6 (March 1953): 10.
197. http://farmworldonline.com/News/ArchiveArticle.asp?newsid=11819.

198. Personal communication.
199. Purdue Agriculturist 47, no. 1 (October 1955): 4.
200. Purdue Agriculturist 47, no. 8 (May 1956): 6.
201. Purdue Agriculturist 47, no. 1 (October 1955): 4.
202. Ibid.
203. Personal communication, August 12, 2014.
204. Personal communication, August 13, 2014.
205. Purdue Agriculturist 49, no. 2 (November 1957): 10.
206. http://memorialwebsites.legacy.com/robertstallwood/memorialSite.aspx.
207. Purdue Agriculturist 49, no. 2 (November 1957): 13.
208. Purdue Debris (1959): 566.
209. htttp://www.oakdaleleader.com/archives/12986.
210. D. R. Frazier, letter to Dean Freeman, September 18, 1959.
211. Ibid.
212. Ibid.
213. Ibid.
214. The Purdue Agriculturist 52, no. 5 (February 1961): 4.
215. Personal communication with Myron Bozell, January 20, 2015.
216. Ibid.
217. Ibid.
218. Ibid.
219. Ibid.
220. Ibid.

Appendix 3. Winter Course Graduates

1. Purdue Agriculturist III, no. 3 (December 1908): 26–27.
2. Purdue Agriculturist XIX, no. 7 (April 1925): 149.
3. Purdue Agriculturist XXI, no. 4 (February 1927): 110.
4. Purdue Agriculturist 47, no. 6 (March 1956): 10.
5. 1943 Winter Course Bulletin.
6. Bulletin of Purdue University Annual Report XLIII, no. 1 (January 1943): 54.
7. Bulletin of Purdue University Annual Report XLIV, no. 2 (February 1944): 35.
8. Bulletin of Purdue University Annual Report XLV, no. 1 (January 1945): 105.

9. *Bulletin of Purdue University Annual Report* XLVI, no.1 (January 1946): 94.
10. *Purdue Agriculturist* 37, no. 9 (May 1946): 9.
11. *Purdue Agriculturist* 38, no. 7 (April 1947): 7.
12. *Purdue Agriculturist* 40, no. 6 (March 1949): 10.
13. *Purdue Agriculturist* 48, no. 4 (February 1957): 4.

Index

C

N

O

P

Q

R

S

T

U

About the Author

John M. Cleland taught biology at Zionsville High School for 30 years until he retired in 2011. Indiana dairy history has been an integral part of his life ever since 1972 when a Purdue Animal Sciences professor gave him two milk bottles. Graduating in Animal Sciences at Purdue University in 1973, John worked in agribusiness before redirecting his life into teaching and coaching at the high school level. However, his interest in and knowledge of milk bottles and dairy history continued to grow. Today John is on the board of directors for the National Association of Milk Bottle Collectors and has one of the largest collections in Indiana. He has had several articles on Indiana dairy history published and has written three books on collecting Indiana milk bottles. In 2014, he was presented a Purdue Animal Sciences Distinguished Alumni award. John and his wife, Judy, have been married since 1973 and live in Indianapolis. They have three children.